G000129476

Preparing and Delivering Effective Technical Presentations

Second Edition

For a listing of recent titles in the *Artech House Technology Management and Professional Development Library,* turn to the back of this book.

Preparing and Delivering Effective Technical Presentations

Second Edition

David Adamy

Artech House
Boston • London
www.artechhouse.com

Library of Congress Cataloging-in-Publication Data
Adamy, David.
Preparing and delivering effective technical presentations / David Adamy.—2nd ed.
p. cm.—(Artech House technology management and professional development library)
Includes bibliographical references and index.
ISBN 1-58053-017-6 (alk. paper)
1. Communication of technical information. I. Title. II. Series.

T10.5. A33 2000 00-039409
601'.4—dc21 CIP

British Library Cataloguing in Publication Data
Adamy, David
Preparing and delivering effective technical presentations.
— 2nd ed. —(Artech House technology management and professional development library)
1. Communication of technical information
I. Title
601.4

ISBN 1-58053-017-6

Cover design by Igor Valdman

685 Canton Street
Norwood, MA 02062

International Standard Book Number: 1-58053-017-6
Library of Congress Catalog Card Number: 00-039409

10 9 8 7 6 5 4 3 2 1

To Georgeanne—my wife, my life partner, and my best friend.

Contents

Preface

THE TECHNOLOGY TO SUPPORT the briefer has grown immensely in the last 10 years. As this goes to press, several excellent slide-generation applications are on the market, and direct projection of slides from computer files is now practical in many circumstances. With all of this new gadgetry, it is tempting to believe that excellent technical presentations are now created by the computer—with little or no effort required by the presenter. This, like many of our fantasies about the life of leisure modern gadgets will provide, is not true. It is still possible (in fact, much too common) to create a presentation with the following problems:

- Visual aids in luxurious color combinations that make them impossible to read;
- Visual aids that are too busy to read . . . but in glorious color;
- Too many visual aids for the allocated time;

- Excellent visual aid format . . . but no content worthwhile to the audience;
- Excellent visual aids . . . poorly presented;
- No discernable focus;
- Technical excellence of presentation . . . but doesn't meet the needs of the briefer or the briefed.

The point is that despite all of that wonderful new software, you still have to plan, develop, and present a briefing properly if it is to be effective. Thus, most of the techniques presented in the first edition of this book are still valid. Only the actual creation and display of visual aids have been impacted by the new technology.

The second edition has new material to deal with the new technology. Since the available presentation software and hardware is changing so quickly, it is impractical for a book of this type to deal with the features of any specific program or piece of equipment. The emphasis is, therefore, on how to use the features available from any of them to the best advantage. This includes "tricks of the trade" to get the most out of the new presentation software and important considerations in the effective use of new presentation hardware.

There are some tricks of the trade that allow you to use presentation software with a maximum of efficiency and a minimum of frustration. Consider the following examples:

- Use of a grid to keep everything lined up;
- Making a solid object of arbitrary shape;
- Tabbing versus spacing;
- Controlling line density on graphs;
- Backup strategies;
- Clip art (why and why not);
- Word-processing tables versus spreadsheets;
- Coordinating (and confusing) slides and handouts.

There are also some important new considerations caused by the newly available presentation hardware. Examples are listed as follows:

- Direct computer projection (when and when not);
- How to convert computer files to old-fashioned presentation media;
- Software compatibility issues;
- Storage media selection;
- Slide transitions in direct projection;
- Using those tiny laser pointers;
- What to do when the infernal machine doesn't work;
- Keeping track of your progress through the briefing.

Properly used, the new presentation hardware and software can significantly improve the audience's perception of a well-planned, -developed, and -presented briefing. However, like an inexperienced driver in a high-powered car, the new gadgets cannot fix a briefing that is not well planned, developed, and presented; they may even make it worse.

Some technology other than your own computer is now available, for example, having 35-mm slides produced (commercially) from computer files, capturing graphics with a scanner, and storing many visual aids on high density storage media. The emphasis is not on the technology itself, but on how to use it effectively (and cost-effectively).

This edition also fills some small holes in the coverage of the first edition, including additional rehearsal techniques, briefing in a language that is not your native tongue (with no translator), and care of your feet and throat when you must give a long briefing or class.

Finally, the author has spent the 12 years since the publication of the first edition presenting multiple-day technical short courses and many other technical presentations all over the world. He has also run several large international technical symposia with multiple speakers. It is impossible to spend that much time presenting and managing presentations without encountering many learning situations. Sometimes innovative techniques worked well, and sometimes they went onto the "I'll never do that again!" list. Comments and helpful suggestions based on both kinds of lessons have been added in the appropriate places throughout this edition.

The following generalities are just as true now as they were 12 years ago:

- Effective presenters are made, not born—you can be in that number.
- An excellent presentation is still an excellent presentation—no amount of new technology can replace attention to the basics of presentation technique.
- Becoming an effective presenter is one of the most dependable ways to enhance your career advancement.

1

Introduction

When their careers start to progress, professionals in any field find that they can no longer enjoy the comfortable anonymity of merely doing good work. At some point, they have to start telling people about it—in writing and in person. The greater the success, the greater percentage of this telling needs to be done in person. In technical fields, career-related public speaking is principally the technical briefing.

Writing is hard work, but, to many people, public speaking is that and more—it's terrifying. A recent poll found that fear of public speaking ranks higher than fear of dying in a significant percentage of the study group. That is sad, because everyone who sticks with the effort finds that the fear is overrated and that the rewards are worth the effort, in terms of both career growth and personal satisfaction.

You will learn techniques from this book to make your technical briefings more effective, so that they will be more informative and immensely less sleep-inducing than most you have heard. However, no mere technique will make you an excellent technical briefer. That will come only with attitude, insight, and practice. This book can help with the first two, but practice is your responsibility.

1.1 The main points of the book

If you get nothing else from this book, please remember these three statements:

1. Good public speakers, including technical briefers, are made, not born. Conscientious practice really does make perfect.
2. If you respect your audience, its members will always be a group of friendly people who want you to do a good job.
3. Excellent briefings are those that tell the people in the audience what they came to learn, in language they can understand, with visual aids they can read.

If taken to heart, these principles will make you a better technical briefer, even if you do nothing else right. The rest of this book is designed to help you deal with the imperatives in these statements.

1.2 About public speaking

Good technical briefing is, above all, good speaking, but it has the added requirement of effectively transferring technical knowledge to the listener. Here are some generalities about public speaking that will be detailed in later chapters.

1.2.1 Handling fear

Even the most experienced professional public speakers are nervous about speaking. Most say that if they lost this nervousness, their performance would suffer. They say that it gives them the edge they need to generate excitement in their audiences. Nervousness is good and natural and probably unavoidable, but fear is bad. Fear can also be overcome.

Think about the reason you fear public speaking—because you fear that you will make a fool of yourself or that the audience will reject you. Feeling this way may well stem from your first public speaking experience, reciting for the class in about fifth grade. The class (probably including you) was made up of a bunch of little savages who gloried in the failure of the one on the spot, forgetting about their own unspeakable misery when they were in the same position. Well, forget it, folks! Unless you are a fifth-grade math teacher, you are probably speaking to a group of mature adults just like you.

Think of how you feel when you are a member of the audience. You are typically just glad that it is not you standing up there talking. You have great sympathy for the speaker (unless the speaker is putting you to sleep or talking down to you). If the speaker is embarrassed, you feel bad for the speaker—you've lost the savage, fifth-grade glee that the person on the spot might fail.

There are, of course, circumstances when a group of adults is more like a fifth-grade class, and those will be dealt with in Chapter 8. However, that is such a rare situation and so easily handled by someone who has developed self-confidence as a speaker that it does not rate much attention.

Remember, if you respect your audience, it will be a group of friendly people who want you to do a good job. If you basically know what you are talking about, prepare well, and think of your audience as friends, you have no reason to fear speaking in public.

You are not convinced, are you? No one who is not an experienced speaker is convinced by that argument, but that does not diminish its truth. Only by giving your first presentation will you become convinced that you will not die from the experience. As you continue giving presentations, you will discover this truth for yourself.

1.2.2 Respect for your audience

Remember, all of the above arguments are based on the assumption that you respect your audience. If you do not show respect for your audience, there is almost no way you can be a successful speaker in any situation—including the classroom.

Everyone has known a professor who "talked down" to the class—we loathed the jerk and sat there only because we would flunk if we missed class! If we learned anything in the class, it was accidental. The same is true of the technical briefer who does not respect the audience.

If you respect your audience, it will show in your eyes, in your voice, and in your body language, and the audience will return the respect by being the friendly people of whom we spoke. They will also open their minds to your message and keep them open as long as you display a reasonable grasp of audience dynamics.

1.2.3 Audience dynamics

Any audience consists of people. If you respect them, they are friendly people, but they are still people. *Audience dynamics* are generalities about the way groups of people act when they are listening to a talk, regardless of the type of talk or the subject matter being covered. The audience-dynamics characteristics discussed here deal with the following:

- Perceived value of talk;
- Interest span;
- Audience reactions versus group size;
- Situation variant factors.

Perceived value of talk

The people in the audience expect to receive something of value in exchange for the time they have invested in listening to you talk, and they arrive for the briefing with some idea of how they will benefit. If they came to be entertained, they expect to laugh or cry or both and will have little patience with a sermon on thermodynamics—and the effective entertainer would not try to deliver one.

More to the point, if the audience came to learn whether or not a piece of equipment will burn up the first time it is used, they still don't want a sermon on thermodynamics. They just want to learn whether or not the equipment will burn. The excellent technical briefer will give them that information with only enough thermodynamic theory to convince them that the answer is true. Giving any additional theory is a waste of the audience's time.

The measure of the perceived value of a talk is an evaluation, from the audience's point of view, of how much value was received per unit of time invested. While this value per time ratio is not a number that can be calculated, it is nonetheless very real to the audience, and the effective briefer gives it strong consideration.

Human interest span

The interest span is the amount of time that a human will willingly continue a single intellectual activity. After that period of time, the person's mind wanders unless something happens to keep the interest going. Educated adult humans theoretically have an interest span of about seven minutes. Therefore, it is a good idea to structure a technical talk so that there is some sort of major transition about every five minutes.

This does not *guarantee* that if a speaker makes a transition every five minutes the audience will stay interested. Such an idea is far from the truth. If your listeners perceive no value from the talk, their eyes will start to glaze over after only a few seconds. This will occur also when the audience is having difficulty seeing visual aids or hearing the speaker. You will have the audience's undivided attention for long periods, however, if your listeners are excited about what they are hearing, because listening will be an emotional as well as an intellectual experience. Take care to keep your talk interesting by showing that the subject is important to you personally, focusing on what is important to the audience, and making some sort of break in continuity about every five minutes.

Audience reaction versus group size

Something to remember about audiences is that the reaction of the audience as a group will be different from the reaction of each individual. Individuals, in one-on-one conversations, react and can change directions quite rapidly. Groups of people react more slowly. The larger the group, the slower the reaction. The feeling associated with speaking to a large group is much like that of moving a large mass on well-lubricated wheels. This is easily illustrated by the reaction of a large group to humor. The larger the group, the longer it will take them to start laughing at a joke—but they will laugh longer.

For example, while speaking to a group of 2,000 people, I told a joke. They took several terrifying seconds to start laughing—I had decided that the joke had died and opened my mouth to continue—but then they laughed for 15 or 20 seconds. Accordingly, the subtlety of your material and the speed with which you change direction must be reduced in talks to larger groups.

Situation variant factors

Another important characteristic of audiences is that they are different in different circumstances. The same audience that was ready to accept deep technical details at 10:00 A.M. will need to be approached more carefully right after a big lunch, because audience members' brains have donated blood to their stomachs to digest lunch. In a restaurant after a dinner with wine, the same audience will need to be significantly entertained in order to accept any technical detail at all—even if they are Nobel laureates.

A final word about audience dynamics

There is an acronym that should be the first law of technical public speaking—BABE, or the Brain can Absorb only what the Backside can Endure. In other words, realize that no matter how interesting your speech is, your audience has physical limitations that must be considered.

1.2.4 A good talk is a good talk

There are many similarities between technical public speaking and non-technical public speaking. Both deal with the transfer of information from a human speaker with a purpose to a group of human listeners who are listening for a purpose. For a good talk to take place, the purposes must be compatible, and the data path must be free from interference. The speaker must be able to be heard and must send the right information. The listener must be able to receive and must be ready and able to accept the information. Otherwise, communication cannot take place.

Technical presenters, particularly engineers, often ignore this, thinking that the sheer brilliance of the technical content of the talk will overcome the obstacles presented by the shortcomings of the humans who are listening. *It is not so, friends!*

1.3 How to do it wrong

The following specific mistakes are often made by technical briefers. They diminish any talk's value to the audience, even when important and accurate information is presented. You will notice that none of them has to do with a shortage of technical expertise by the speaker.

- Failure to set limits on the subject;
- Failure to plan for proper use of the time available;
- Overcomplexity of visual aids;
- Use of terms not familiar to the audience;
- Lack of organization of the presentation:
- Failure to prepare properly for the presentation itself;
- Failure to set goals for the briefing.

Ways to correct each of these speaking problems will be dealt with in detail in the rest of the book—not just what to do, but how to do it.

1.4 Outline of the book

This book is organized into chapters that are grouped into general areas of concern, some of which are complex enough to have required multiple chapters. The areas of concern are listed as follows.

- *Designing the briefing:* Including goal setting, scope limitation, organization, and consideration of audience dynamics;
- *Using visual aids:* Including the jobs visual aids need to do, the strengths and weaknesses of various visual aid media, how to fit visuals into the overall talk, ways to organize and limit the material on individual slides, and ways to present visual aids optimally;
- *Designing and producing visual aids:* Including slide layout, choice of letter and graphic element sizes and colors, projection of unity, and production of all types of visual aids with or without a professional graphic arts department (with a special emphasis on the production of professional looking graphics by making maximum use of new presentation development software);

- *Briefing room logistics:* Including room layout, handling of visual aids, and "care and feeding" of the audience;
- *Presentation technique:* Including speaking mechanics, using visuals, handling distractions, and answering questions;
- *Management of briefings by multiple speakers:* Including design of the entire set of briefings, assignment of tasks, selection of speakers, unity considerations, management of speakers, and the mechanics of conducting the full briefing;
- *Real-world considerations:* Including practical help in preparing and presenting the most common types of briefings required of technical professionals and tips for handling challenging speaking situations.

2

Before You Even Start

THE FIRST MISTAKE frequently made by technical briefers is starting the preparation process at a point far beyond where it should start. Most technical professionals assume (usually correctly) that they have been chosen (or "sentenced" as the case may be) to give the briefing because they know more about the subject than any other available person. With this knowledge, they start by jumping right into the technical details of what they know about the subject of the briefing. This kind of a start not only makes the preparation process longer and much more painful than it needs to be, but also makes it difficult to give an effective technical briefing when the process is finished. Starting with the details always leads to a reiterative process of removing extraneous information and going back to add more detail after you realize that the whole briefing doesn't "get the job done." In the preparation of an effective briefing there are several important steps that will streamline the process before you even get near the technical detail.

2.1 The preparation process

Figure 2.1 shows the entire process of preparing an effective technical briefing. Every step is required to produce a quality product. The initial items are 10 important tasks that should start the process. Depending on the subject and the situation, these might take only five minutes or they might take much longer, but they should never be ignored. The design of the talk is the development of a structure for the talk that will control the way information will be organized and presented. Detailed preparation involves the incorporation of the technical data into the structure developed during the design phase. Finally, some level of rehearsal (depending on the formality of the briefing) is a part of the preparation of any effective presentation.

2.2 Initial items

Before you touch pen to paper (or fingers to computer keys) there are 10 important tasks that will aim your preparation toward success. Skip them at your peril!

1. Determine the *purpose* of the briefing. (What do you want the audience to *do* as a result of your talk?)

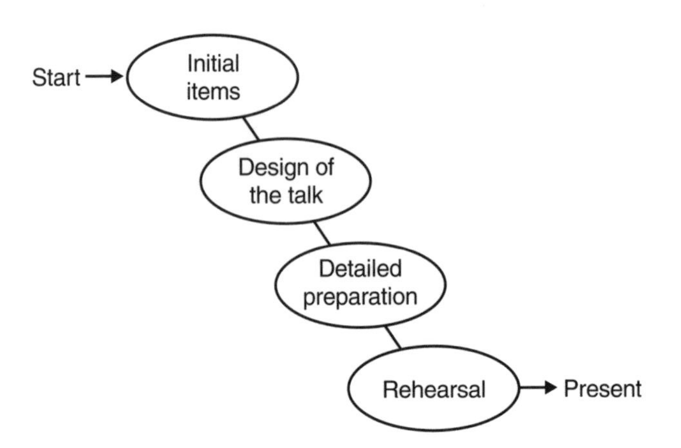

Figure 2.1 The preparation process.

2. Decide what the audience needs to *understand* to take the action in task 1.
3. Decide what *two* or *three* things your audience should *remember* to take the action in task 1.
4. Find out *who* the audience will be.
5. Find out *what* the audience *knows* about the subject.
6. Try to figure out what the audience *expects* from your presentation.
7. Find out how much *time* you have for the talk and questions and answers.
8. Find out the *size* of the audience.
9. Find out *when* and *where* the briefing is to take place.
10. Find out *what else* the audience will be *doing* before and after your briefing.

Figure 2.2 shows these initial items in relation to the rest of the preparation process. Through these 10 tasks, you will accumulate information on the purpose, the audience, and the logistics. Then you will have the rest of the information you need (in addition to what you know about the subject) to design and prepare an effective presentation.

The person who asked you to present the briefing will probably be the best source for the answers to these 10 questions and may be able to supply all of the answers offhand. If not, take the trouble to find out who can answer your questions and gather the data. This information is as important to the success of your talk as your expertise in the technical field and your knowledge of the subject. Now, we will deal with each of these 10 initial items in more detail.

2.2.1 The purpose of the talk

The first thing to do when you have been asked to give a briefing is to figure out the *real* reason you are giving the briefing. Incidentally, the real reason is not just to convey technical information to your audience; it is to

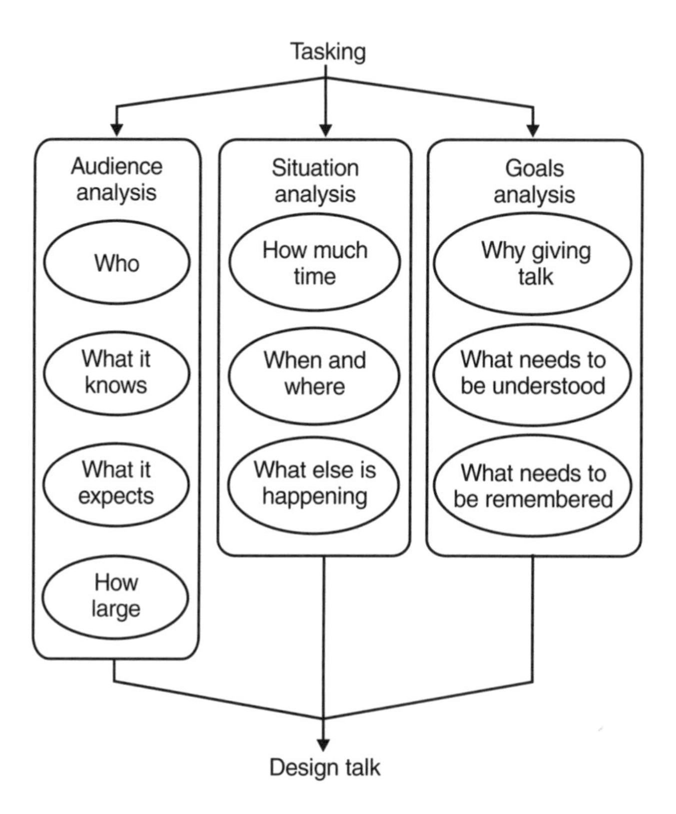

Figure 2.2 Initial items.

cause your audience to take some specific action as a result of your having spoken. Every talk causes the audience to take some action.

"Not me," said the red-eyed conference participant, "he just put me to sleep."

Sorry, that *was* an action—probably not the action that the speaker or the conference planner had in mind (unless it was a conference on hypnotism)—but an action all the same.

The difference between an effective technical presentation and an ineffective presentation is that the effective speaker causes the audience to take the *desired* action. The effective speaker also takes care to decide ahead of time exactly what action is wanted and plans the talk to induce that action effectively.

The desired audience action can be anything from the above-mentioned sleep to a full-blown riot, but for technical talks the action typically falls into one of the following categories:

- A decision to buy something;
- Agreement with the speaker on some important issue;
- Understanding of some concept;
- Increased confidence that something will be right.

To bring this into better focus, consider some specific examples. The decision to buy something hides under many disguises:

- Indeed, you may be trying to get a customer to buy some product.
- You may not be trying to close a sale, but your talk describing a piece of new software may well be the key element in a large sale for the company.
- The thing being sold may not be a product at all, but a concept. For example, you may want the organization to adopt a new procedure that will make your department more efficient.
- You may want to convince the boss to sign the cover letter of the proposal you are presenting for final management review.

All of these are selling situations, and by honestly stating the real purpose of the talk at the very beginning of the preparation process, you can properly focus all of your efforts on making the sale.

Trying to achieve agreement is essentially the same as selling something, but it is helpful to consider it separately because the intensity of action is usually different and the approaches may be different. The goals are typically more limited than in out-and-out selling, because when selling, the action is to get the sale—or the boss's signature on the dotted line—which is a commitment to a single significant action. Achieving agreement does not require that kind of final action, just movement of opinion in the desired direction. The following are examples:

- The audience *knows* that a concept will never work; your job is to convince audience members that there is a reasonable chance that it might.
- A group of customers needs to be convinced that your company is a "good corporate citizen."
- The big boss is in town, and you have the job of describing your department, but the real job is to cause him or her to have a very high opinion of the department.

Causing the audience to understand a concept is the talk purpose with which technically oriented people are most comfortable, yet it is probably the purpose that is most often left unfulfilled. The problem results from the confusion between *explaining a concept* and *causing an audience to understand a concept*. To explain a concept, you need only know the subject, open your mouth, and talk about it. Causing the audience to understand that concept requires not only knowledge of the subject, but also knowledge of the audience's capabilities or background and a way to help the audience grasp the new knowledge about the subject. Examples of talks with this purpose are listed as follows:

- A lecture to students in a classroom situation;
- A briefing to management on a new concept or product;
- An after-sale briefing to customers on how to use something they have just purchased;
- An explanation to new employees of job procedures.

In each of these types of talks, the proof in the pudding is whether or not the audience understands what you have been saying when you are finished.

Giving the audience increased confidence that something will be right is the classical purpose of the project review, although it also fits many other types of talks. You have been doing a job and now it is time to show and tell. The reason that project reviews are conducted is to keep the big boss from being surprised if a big, multifaceted project does not come together at the end. Even more to the point, project reviews allow the boss to take direct action to correct anything that is found to be wrong along the way. Since you do not want to be one of the "things that are corrected," it is important that the big boss understand that your part of the job is on schedule, on budget, and on course to meet all of its technical objectives.

2.2.2 What does the audience need to understand?

After you establish the purpose of your talk, decide what the audience needs to understand to take the action you want it to take. The reason this is so important is that the audience will help you limit your subject. Since you will have a limited amount of time to present your talk, you want to be certain to move your audience toward the desired decision or action. Any information you present beyond that required to support the desired action is a waste of time. You will see that a few minutes spent on this point will pay big dividends in effectiveness.

To accommodate the audience's needs effectively, put on your "audience hat" and consider what your thoughts would be if you were being convinced to take the same, desired action. If your purpose is to sell something, why would you buy it? If your purpose is to convince, what would it take to convince you? If your purpose is to cause understanding, what points would you need to focus on to grasp the overall concept that is being presented?

For example, if you are selling, consider the reasons you would buy. The old saw about selling is that you should push the benefits not the features of the item being sold. How will you, the customer, be happier or healthier or wealthier if you purchase the product? The following are typical reasons for buying:

1. It is cheaper than the competition (if it is not, do not bring it up).
2. It will allow the customer to perform a process in a more efficient way, thereby saving enough employee time to more than cover the cost of the new purchase.
3. It will require little or no maintenance.

4. It will do the job better than anything else on the market (thereby making the customer's customer or boss happy).

If you are convincing, perhaps you can consider the standard "major premise, minor premise, conclusion." If the process is more complex, write down the individual concepts the audience *must* believe to be convinced that the overall premise is true. If you are teaching a group how to do something, carefully consider the key procedural steps—the steps you want them to master.

2.2.3 What the audience should remember

Think back to the last time you heard a talk of any type—a technical briefing, a lecture, a sermon, or any other. Now, try to remember something about the material presented. If you are like most people, within a few minutes after the talk you were able to recall only two or three things that were said. It may be the joke that was used to open the talk. It may be a single important point that was made repeatedly. It may be the bump you got when your head hit the table after you fell asleep.

You may be one of those rare individuals who have photographic memories and remember everything. Generally, the rest of us mere mortals will remember only two or three items, no matter how long the talk lasted. Quantities of detailed information can be carried from a meeting in written form or studied from a book beforehand. Such detailed information can certainly be affirmed and enhanced by the spoken word, but most likely, the lasting effect of the talk itself is the retention by the audience of only two or three items. What a shame if the only thing remembered from your talk is the punch line of the nonrelated joke you told before getting into your subject matter.

Since the audience will remember only two or three items, it is easy to see the value in carefully planning what those two or three remembered items will be. Once you have decided on the items, you can use various devices and schemes (discussed in detail in later chapters) to be sure that the audience remembers them.

For example, pretend that you are making a talk that is basically a sales pitch. You have determined that the audience needs to understand five or six benefits from the product to decide to buy it (low cost, low maintenance, adequate performance, etc.). However, the most critical *new* information you are bringing them is that a flaw in the design, which kept them from buying it last time, has now been corrected. With this in mind, you want them to wake up tomorrow morning thinking the following:

- The design has changed.
- Independent testing has proven that the defect is now gone.

2.2.4 Who will the audience be?

If you can find out exactly who is to be in the audience and if you know all of those people, you are most of the way there. Unfortunately, there is usually a large number of strangers in the audience, a critical mass about whom you will need to gather the best information available. You will do a much better job of preparing the briefing if you know at least the following about audience members:

- Their technical education level;
- Their seniority in their field;
- Their related real-world experience;
- The prejudices they bring to the occasion;
- Their official functions (customers, bosses, etc.).

You need to know their education level to be able to judge the appropriate level of vocabulary to use. It would be a disaster to explain a concept using differential equations to an audience of managers who have had no math since high school. On the other hand, if they have their Ph.D.s in mathematics, you certainly do not want to explain what an integral sign means.

You need to consider how senior the members of your audience are in the field to know what general background information they will have. Do they know the industry? Do they know the competition? If you drop the names of the old masters in the field, will they think you are talking about rock music stars? Do they know the inside jokes in the general field? Do they know the buzzwords?

You need to know about their real-world experience to determine what it will take to impress them. For example, if they have never been in a military aircraft, they'll probably not be impressed with the high level of vibration that must be survived by hardware mounted in such aircraft.

As human beings, we carry around many more prejudices than those concerning race, sex, and others that are addressed by civil-rights legislation. A prejudice is any prejudgment that a human being has made on any subject. "My mind is made up, don't confuse me with facts," is a very human statement. Not all prejudices, of course, are bad. The "prejudices" (also called instincts) of a senior-level professional in any field typically allow the old pro to be much more effective than a well-educated but new professional in same field, even in a situation that the old pro has never encountered. Whether you share the prejudices of your audience or not and whether you plan to try to change a prejudice or not, you will give a much more effective talk if you spend a few moments considering the likely preformed opinions of your audience members.

Discretion dictates that you know the official functions of the members of your audience. If there are competitors present, you will, of course, want to avoid spilling the company's secrets.

Your talk becomes a serious legal problem if it divulges government security information or transfers technology to citizens of other countries. Some information is restricted by law from release to unauthorized people. If any type of restricted information is involved, be sure that you know the official status of the audience and adhere to the rules and guidance from the appropriate authorities. Otherwise you might face criminal prosecution.

2.2.5 What the audience knows about the subject

Your concern about your audience's knowledge of the subject extends beyond their level of general expertise. You want to know how much the audience knows about the specific subject of the presentation. Whether you are selling, convincing, or explaining, you do not want to be halfway through your talk dealing with the fine points of trading stocks and bonds and be interrupted by a question about what a "Wall Street" is.

Consider the following true story:

> The manager of an organization that manufactured state-of-the-art electronics needed to hire some experienced "trimmers" to handle laser devices used to trim microscopic amounts of material from electronic circuits. The job required some specialized education and several weeks of training. The manager passed this requirement on to the personnel department. Assuming that they were familiar with the operation, the manager didn't waste their valuable time with all of the details. The first candidates brought in for interviews were "experienced trimmers" from the local cauliflower sheds.

While it is important that your audience have sufficient background information, it is a waste of valuable time to explain something that everyone knows. When in doubt, of course, the speaker should err on the side of explaining too much rather than too little.

2.2.6 What the audience expects

The audience may come without any preestablished expectations about your talk, but if they do expect something (for example, when a meeting agenda has been distributed), you need to know about this early in your preparation.

Audiences are much happier when given what they expect. However, when there is a good reason for you to talk about something else, or if you are not the speaker they expected, or you do not have the credentials they were led to expect, you will need to correct the expectation as part of your introduction or handle the problem in some creative way to remain credible to the audience.

2.2.7 How much time is allowed?

You need to know how much time is allowed for your presentation as well as the amount of time allotted for a question-and-answer period. This will help you limit your subject and the design of your visual aids.

2.2.8 Audience size

The size of the audience has some straightforward logistical implications, such as the necessity for a microphone and the appropriate choice of visual aid media. It also has some implications that are not so obvious—the larger the group, the less homogeneous it will be and the more formal the presentation should be.

2.2.9 Time and location

It does seem obvious that you should know when and where you are to speak, but do not take it for granted; find out when you are first asked. The amount of time available for preparation and the facilities available in the briefing room will dictate your choice of visual aids.

2.2.10 What else is happening?

You will not be giving your briefing in a vacuum. Other events in your audience's routine will affect its members' reaction to your talk and should become a part of your preparation process. Are you the first speaker in a full day of briefings? Are you the last? If you are on right after a big lunch and decide to darken the room for a film, you will need to plan on making a loud noise to awaken the group before you turn the lights back on.

To plan your own briefing properly, you should know what is to be covered by other speakers and the overall goal of the program.

2.3 Designing your talk

At this point in the preparation process, you have gathered much information that has little to do with the subject matter of the talk you will give. The assumption has been that you know what you are talking about in the subject field for the briefing. (If not, this might be a very good time to learn.) Congratulations, you are now ready to design the talk. Figure 2.3 shows the steps in the talk design, a process that is distinct from the task of actually preparing the talk. Like the initial items, the design of the talk can take but five minutes or a month of serious work. This will depend on the nature and scope of the talk, but it should always be done before moving on. Figuring out where you are going and how you will get there will make the trip more efficient.

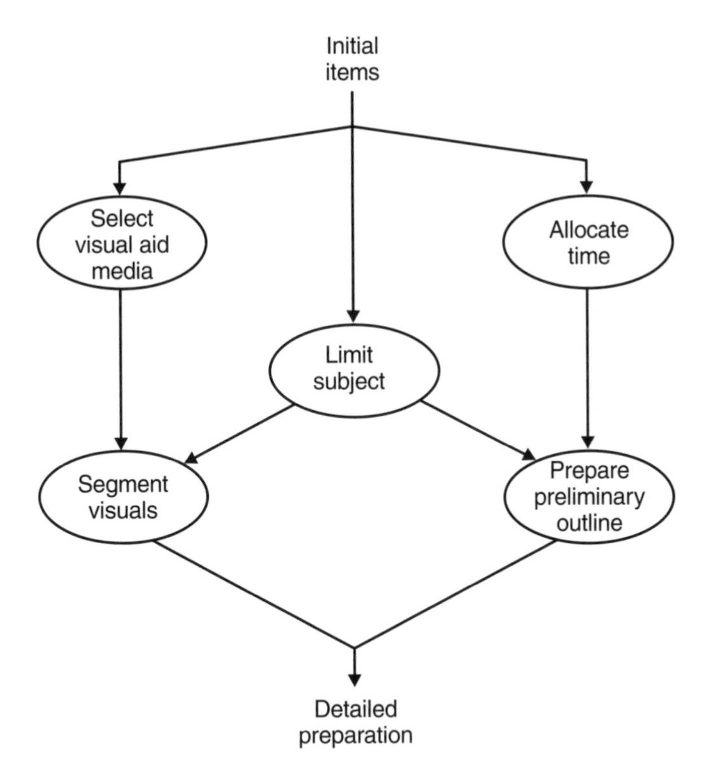

Figure 2.3 Designing the talk.

As shown in Figure 2.3, there are three independent tasks that start the design process. These are the selection of visual aid media, limitation of the subject, and allocation of time. Once these are done, the preliminary outline and segmenting of visuals will complete the design so you can prepare the talk content.

2.3.1 Selection of visual aid media

About 80% of the information humans receive comes in through their eyes. For technical information, that percentage is probably higher. Therefore, in a technical talk, which must get technical information into the brains of the audience to be effective, the selection of a visual aid medium may be the most important choice made by the speaker. Most of the time you will have to select your visual aid medium from the two or three choices that are "always used" by the organization that owns the room in which you will speak. Frequently, the choices are not as restricted as they first appear. Equipment can usually be borrowed, rented, or improvised if necessary.

The proper visual aid medium can make a significant difference in the effectiveness of your talk. The choice of a visual aid medium should be made based on audience size, room layout, and the nature of the material being covered. In Chapter 4, the advantages and proper applications of various visual aid media are discussed in detail.

2.3.2 Limitation of the subject

Every subject contains more information than can be covered in a talk of *any* length. You are well-advised to eliminate all information that does not fall into the category of what the audience needs to know to take the action you desire. Eliminate it immediately. Any additional subject matter will diminish the effectiveness of your talk by diverting the audience's attention from the subjects you want its members to consider.

Obviously, the scope of the subject covered is also limited by the time available, but it is important to understand that this is a secondary matter. If your purpose requires the audience to have more information than you can transfer in the time available for your talk, you are in a classic "plan to fail" situation and need to either reduce the scope of your objective or get more time. However, this is seldom the case. After a careful analysis of what your audience really needs to know to take the desired action, you can normally reduce your material to fit into the time available.

If you have extra time, it is better to cover the necessary information in more depth rather than to add to that information. It is a very rare speaking situation, however, in which anyone minds if the speaker quits a little sooner than planned. The moment when you have finished saying what you have to say is a very good time to quit. As you know from experience, the world already has enough speakers who run out of material before they run out of wind.

2.3.3 Allocation of time

Whether you are giving a five-minute presentation or a five-day seminar, it is important to allocate carefully the precious time that the audience is available to you. The proper allocation of time is subjective and dependent on the nature of the subject matter and the level of the audience, but a few rules of thumb can be applied.

First, consider time for questions and answers at the end of your formal presentation, if a question session is part of the format of the speaking occasion. A good rule of thumb is to allow 20% of your available time for questions from the audience. This percentage may decrease for a longer talk and increase for a shorter talk.

If the subject matter is new to the audience, you might expect more questions. In real life, however, you will probably experience fewer questions because the members of the audience will take what you say at face value and because they are not comfortable in phrasing questions. If the listeners know quite a bit about the subject, expect many questions, since they are comfortable with the vocabulary of the field and will be sharpening the fine points of their understanding.

In the formal part of your talk, you will need to divide your time among the introduction, the body, and the conclusion. In every effective talk of more than one sentence in length, all three will be present. Much more will be said about these in Chapter 3.

The rule of thumb for the time division within your formal talk is 10% each for the introduction and the conclusion. However, in a short talk (less than 15 minutes) 20% for each of these parts is more appropriate.

By combining these two allocation numbers (80% for formal talk and 80% of that for the body), you will note that a *maximum* of two-thirds of your total time is available for the body of your talk, which (as you will see in Chapter 3) is the only part available for the delivery of hard information. This concept is horrifying to many technical speakers, since they are focused on the material rather than on inducing their audience to take the action required.

2.3.4 Preliminary outline

The preliminary outline is not much of an outline. It does not get into the way you want to present the information; it just sets down what you need to cover during the talk. It is important to the process because it encourages you to recheck the limitation of the subject against the time available and to segment your visuals effectively.

2.3.5 Segment the visuals

Before you start creating visual aids, it is important to carefully consider what should be covered on each frame of information displayed to the audience. If more than one visual aid medium is used (for example, videotape and slides or flip chart and blackboard), you will need to decide what information fits best in each medium. Chapter 4 will detail the appropriate use of each type of medium and how much information should go on an individual slide or flip chart page, among other information.

At the end of the segmenting process, you should have a list of individual visuals (for example, slides). Each visual should have a working title that defines the subject matter as well as a description of how to present the information. An example of such a list follows:

- Slide 1: Company organization chart;
- Slide 2: Photo of front of building;
- Slides 3–9: Photos of company products;
- Slide 10: Line drawing of laboratory layout.

2.4 Detailed preparation

Once you have completed the initial items and designed the talk, it is finally time to perform the detailed preparation tasks. As shown in Figure 2.4, the detailed preparation of the talk is the process of writing a detailed outline, sketching the actual visual aids, writing any text required, assembling the draft briefing, and preparing the final versions of the visual aids. Each of these activities is covered in detail in Chapters 3, 4, 5, and 6.

2.5 A funny thing happened on the way to Chapter 3

Humor involves very powerful imagery, and it is often misused in technical briefings. In many circles, a joke is expected at the beginning of every technical talk. Contrary to prevailing opinion, the concept of the opening

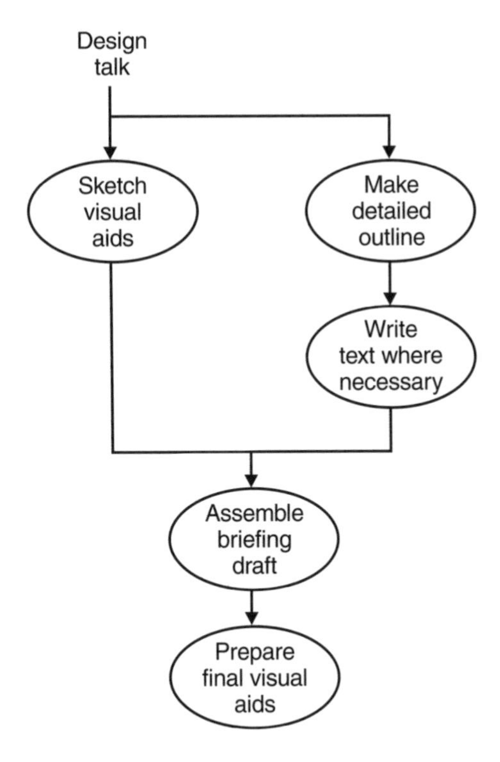

Figure 2.4 The detailed preparation process.

joke did not fall from the sky for the benefit of humankind; it just evolved. Its popularity may come from the perception that most briefings are so boring that a joke plays the role of a last meal for the condemned audience before the torture of the talk ahead.

This seeming requirement for jokes causes a significant problem for speakers who, while they have a great deal of useful information to impart, simply cannot tell jokes. These poor souls typically start by telling a joke so badly that the audience is in pain—or by stammering out weak apologies for not knowing any jokes. Less obvious, but just as ineffective, is the speaker who tells a joke unrelated to the subject at hand. In each case the speaker has lost the impact of an effective opening.

An opening joke is simply not required. If you're not comfortable with humor in this or any other speaking situation, do not use it; several alternative ways of opening a talk are discussed in Section 3.4.

Bear in mind that jokes are not the only form of humor appropriate to public speaking. Often the funniest speakers do not use jokes at all—they just use occasional overstatement or understatement or use an obviously inappropriate word or an unexpected gesture. For example, one speaker had a cardboard cutout of a hand with the index finger extended. The audience thought it was the speaker's hand pointing out items on the overhead projector slide. He walked away leaving the hand on the screen and brought the house down.

If you do use humor, use it to make a point. Remember item three in the "things to do before you start" list. Your audience will probably only remember two or three things that are said. If you have a joke that makes your strongest argument or if you can modify a joke or generate a humorous statement to make that point, you will ensure that the audience remembers what you want them to remember.

Be very careful of off-color jokes. Vulgarity is almost always out of place in public speaking. Even in a room full of drunks, there will probably be someone who is offended by vulgarity, while no one is ever offended by the lack of vulgarity.

Ethnic or other "group" humor (fighter pilots, engineers, men, women, etc.) must be used very carefully if at all. A reasonable rule of thumb is that you can get away with this humor if you are a member of the group and the humor is *mildly* self-deprecating or if you are not a member of the group and the object of the humor "wins." If there is a chance that your use of "group" humor might offend or alienate audience members, you should avoid such humor. Mean-spirited humor is always out of place. If someone in the audience is offended by vulgarity or an ethnic slur, you can be sure that person will remember nothing else you have to say.

3

Outlining

THE OUTLINE IS THE FRAMEWORK of the presentation, on which all of the substance of the talk will be built. There are many acceptable ways to approach the outlining process. Some are more appropriate for some subjects and speaking situations than for others. This chapter will cover the generalities of organizing material within a talk and give some helpful techniques for the outlining process.

Modern presentation development packages can automatically format an outline as you are developing it and can even convert an outline into a series of bullet charts. It is important to understand that the considerations presented in this chapter are still required if your briefing is to be coherent and to accomplish its goals.

3.1 General structure

You have heard it before and will hear it again: Every talk should have an introduction, a body, and a conclusion. Tell them what you are going to tell them, tell them, and then tell them what you told them.

You may think that this is a waste of time. Why not just present the information one time and be done with it? If you were a machine, passing information to another machine, this three-part approach would not be necessary. However, you are not talking to a machine; you are talking to human beings who process data in a significantly more complex and adaptive way than that used by machines. If you have worked with computers, you may doubt this statement from time to time, but machines basically do just what they are told in the way they have been told to do it without personalizing or analyzing the input. People are simply not that way. We decide what information to accept and we apply our own attitudes and perceptions as we evaluate it before allowing the new information to affect our actions. The use of an introduction, body, and conclusion and the finer-scale structure of an effective technical presentation assure optimal data processing by your audience.

3.2 Time allocation

After considering the situation in which you will be speaking, you will need to allocate time to various parts of your talk. The generalities discussed in Chapter 2 and summarized in Table 3.1 should be followed unless there is a pressing reason to vary from them. Examples of circumstances in which different time divisions might apply include the following:

- Talks to a completely new audience when no one is introducing the speaker. In this situation, you must introduce yourself to the group and then introduce the subject, requiring more introduction time.
- Very long talks by a single speaker (for example, a multiple-day seminar)—but be careful in this situation since it may be appropriate to consider the full seminar as a series of one- to two-hour talks, each of which should have an introduction, a body, and a conclusion.

Table 3.1
Allocation of Time During Presentation

Talk duration	Less than 15 min.	More than 15 min.
Introduction	20% of talk time	10% of talk time
Body	60% of talk time	80% of talk time
Conclusion	20% of talk time	10% of talk time
Questions and answers	20% of total time for talk and questions and answers	

3.3 The parts of the talk

Each of the parts of the talk aims to accomplish a specific mission in support of your goal—to cause the audience to take some desired action. In addition, there are certain items of business that are placed properly in the various parts. Sections 3.4–3.6 deal with the three parts of the talk in more detail.

3.4 The introduction

The introduction is your all-important first impression with the audience. It should grab their attention and establish a receptive attitude in support of the rest of your talk. The "things to do list" for the introduction includes the following items:

- Get the talk started.
- Present your credentials.
- Give the audience an overview of what you will present in the body of the talk.

3.4.1 Getting the talk started

We talked about the "obligatory opening joke" in Chapter 2. Perhaps "disparaged" is a better description of what we did to the opening joke, but it must be said in its defense that a joke does get the audience's attention and get the talk started. Contrary to popular opinion, however, this is not the only way to start a talk and is often not the most effective way. Commonly used ways to effectively open a talk include the following:

- Make a strong statement of fact that you plan to support with the rest of your talk. For example:
    - "The design of the power supply completely satisfies all specifications."
    - "Acquisition of a left-handed gerblesnatcher machine would probably double your profits during the next six months."
- Ask a question, especially a question you can answer and that requires only a head movement or a mumbled "yes" or "no" in response. Follow it with a crisp answer. For example:
    - "Is everyone familiar with the problem areas presented last month? Well, they have all been fixed."
    - "Do you know what happens if the power is turned on before the ground is connected? Yes, somebody gets burned . . . maybe you."

  It is best to avoid questions that require some real commitment by the audience during the introduction of your talk (for example, the raising of hands or individual answers), unless you really need the information or somehow have already established excellent rapport with the audience. This caution is given for two reasons: First, the audience will not likely participate in your talk before being "warmed up," and a halfhearted response will diminish the receptiveness of the audience to the rest of your talk. Second, complex answers might evolve into a question-and-answer session before you have started to talk.
- Give a graphic description of something related to the subject. For example:
    - "This morning when we arrived, smoke was pouring from the top of the electronic control box."

    - "Three highway bridges have collapsed during the past 10 years."
    - "Over 1,000 left-handed gerblesnatchers are now in service in 12 countries."
- And yes, you *could* open with a joke as "everyone" does, but keep the following in mind:
    - Be sure it is in good taste and appropriate for this specific audience. (See Chapter 2 about vulgarity and ethnic humor.)
    - Be sure that humor is appropriate ("A funny thing happened on the way to this funeral" does not fly).
    - If you cannot tell a joke, learn how to do so. There are references in the bibliography that include information on how to develop humor in public speaking.
    - Consider saying something in a humorous way (using an overstatement or understatement) in place of a formal joke.
    - Above all, be sure to use the joke to make an important point relative to the *purpose* (not just the subject) of your talk.

3.4.2 Present your credentials

This is not always necessary but must always be considered. One of the strange and wonderful things about human beings is that we decide ahead of time how wide to open our ears. If the speaker has great credibility, we will believe almost anything that is said. If the speaker is not credible, we will reject even the most self-evident truth.

When you are to be introduced by someone else (such as a meeting coordinator or master of ceremonies) it is best to have that person provide your credentials. A listing of your virtues sounds much better coming from someone else's lips. Then you can have the credibility you need to be effective and still humbly dig the toe of your boot into the prairie sod and say, "Aw shucks, ma'am, I ain't all that wonderful." Also, if another person can establish your credibility before you open your mouth, you can use your opening to make a point relative to the objective of your talk.

Experienced speakers plan the way they are to be introduced as they plan the rest of the talk. Be sure that the introducer has the proper information to introduce you in a way that will make you credible to *this* audience on *this* subject at *this* time. If you are to be introduced by a klutz who may foul up the introduction, you can write out a script for him or her.

If there is no one else to do it, you must handle the credibility issue yourself. Even if everyone in the audience knows your complete background, there is still some credibility work to be done—and it must be done in the introduction.

Your general background is not sufficient information. The audience also needs to know that you have some special reason for being considered credible on this subject at this time. Perhaps you have recently conducted research that has developed new information that the audience does not have. That will make you credible even if you are a high-school dropout speaking to an audience of Ph.D.s.

3.4.3 Tell them what you are going to tell them

The reason you "tell them what you are going to tell them" is not that repetition helps people remember. The reason is that people receive information more efficiently if they know what to expect. Time for another true story:

> I was a dinner guest in a German home and was having a delightful time talking with the family children (two boys, aged 8 and 10) while their mother put the finishing touches on the meal. They were speaking fluent German while I was straining the limits of my linguistic ability. The subject for several minutes had been American cowboys, but the younger boy suddenly interjected, "Guess what we're having for dinner." Although I knew all of the words he used, his statement completely stopped the conversation while I desperately tried to fit those words into the subject of American cowboys.

The situation in this story is more analogous to technical briefing than you might imagine, since many in the audience may be stretching the limits of their comprehension of the subject matter and vocabulary you are presenting.

Whether or not your audience is familiar with the subject, you will communicate more effectively if you lay out the general structure of the talk in the introduction. This gives the audience a chance to generate a mental "form" on which they can "fill in the blanks" during the body of your talk. By giving them this forewarning, you make them ready to accept the more detailed explanations and descriptions in the body of your talk. They will waste no time figuring out how the information you are giving applies to the points you are trying to make. Because your audience will remember only two or three things you actually say during your talk (see Section 2.2.3), it is an excellent idea to incorporate the two or three things you want remembered in your introduction.

3.5 The body

The body of the talk is what the audience members think they came to hear. Its function is to supply the hard data that will motivate the audience to take the action you want them to take. It includes supporting facts for the arguments you make in the introduction and the conclusion and all of the detailed information you have for the audience.

The body of the talk should be organized so that it is as easy as possible for the audience to accept the information presented. Unfortunately, many speakers mention facts in whatever order they spring to mind, keeping the audience in a constant state of confusion.

In a well-organized presentation, the information seems to flow naturally from point to point. It seems inevitable that the material will be presented in the chosen sequence.

To achieve the deceptive ease with which experienced briefers move through their material, choose an organizational approach in the process of designing your talk. There are several classical organization structures that are commonly used for various types of talks, and they are discussed in detail in Sections 3.5.1–3.5.6. However, it is important to realize that no organization structure is always right for a specific type of talk. The acid test is whether or not the organization seems natural for the subject at hand.

To achieve a naturally flowing organization of your material, put on your "audience hat" and consider what the most natural flow of material would be, taking into account what you believe the audience knows about the subject. Making a diagram like that shown in Figure 3.1 is frequently helpful in the organizing process.

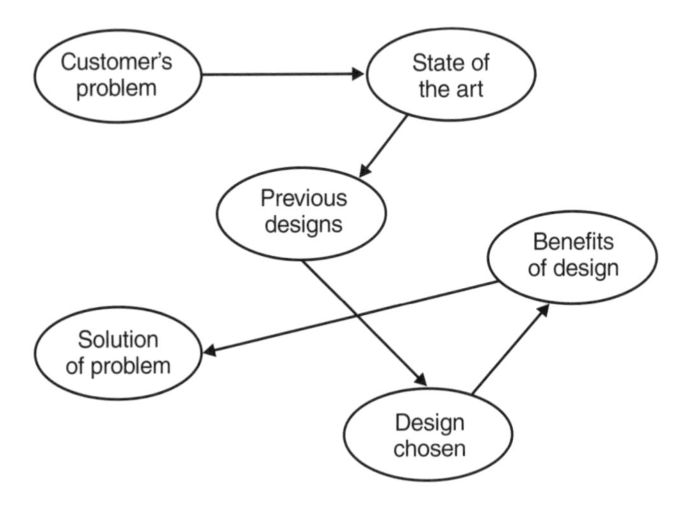

Figure 3.1 Logic diagram for briefing.

With the above advice firmly in place, it is time to consider the classical organization techniques used for various types of technical presentations. The most commonly used are the following:

- General to specific;
- Input to output;
- Stroll through the park;
- Time sequence;
- Background to problem to solution;
- Increasing power of arguments.

This must not be considered a complete list of organizational approaches. Any logical way of covering the subject is acceptable, and in some cases you will need to use combinations of approaches.

3.5.1 General to specific

The general-to-specific organizational technique is most appropriate when the talk requires the explanation of some complex concept. It involves starting with a top-level description of the whole device, process, or concept. Then, as the talk progresses, the big picture is broken up into smaller and smaller pieces until an adequate level of detail has been covered. The big advantage of a general-to-specific organization is that it immediately makes the audience aware of the big picture. With this in mind, the audience will not get lost in the technical details because audience members can incorporate each specific fact into the big picture as it is received.

The general-to-specific technique is so powerful that it is incorporated at some level into almost every effective technical presentation, regardless of the overall organizational technique chosen. It is used for any detailed descriptions that are required. Chapter 4 presents special techniques for handling visual aids in the general-to-specific organized talk.

Do not confuse this with another, related technique that is taught in nontechnical writing courses. That is specific to general, in which all of the specific facts are presented and then incorporated into the big picture at the end. Specific to general is a superb and beloved technique for murder mysteries and spy stories, but it is almost always a disaster when used in a technical briefing. The same gradually evolving big picture that makes people enjoy spy thrillers will put them to sleep or send them screaming from the room if they are trying to grasp technical concepts. *Specific-to-general organization is not recommended for technical briefings.*

3.5.2 Input to output

The input-to-output approach is recommended for the description of any type of processing equipment. If anything from electronic signals to rutabagas goes into a machine and comes out in some other form, the input-to-output approach works well. The technique involves stepping through the processing device in the same way that the processed material passes through it.

Consider, for example, a talk to explain how a commercial FM radio functions. Following the input-to-output technique, the description would follow the path of the received radio signals as they pass through the receiver. Figure 3.2 is a block diagram of the radio. The signals proceed as follows:

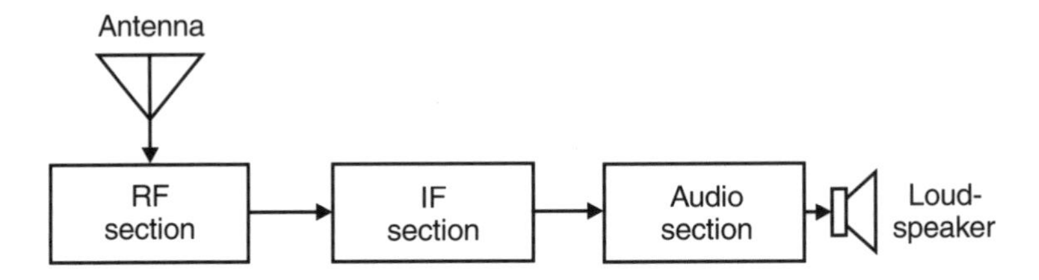

Figure 3.2 FM receiver block diagram.

1. Enter the receiver via the antenna;
2. Are tuned and downconverted in a radio frequency (RF) section;
3. Are amplified and filtered in an intermediate frequency (IF) section;
4. Are converted to audio signals by a discriminator in the IF section;
5. Are amplified in the audio section;
6. Are finally output to human ears via loudspeakers.

First, you describe the antennas, then the RF section, and so on through the system to the loudspeakers. This allows the audience to follow your material easily because they know what to expect.

Some items in the radio are not directly in the signal path (for example, the controls on the front of the cabinet). To cover these items, you make a small side trip at some logical point in the description and then return quickly to the main path. For the example of the controls, it would be logical to cover them just after you have described the tuning process in the RF section. Then, you can return to the signal path for the IF section description.

3.5.3 Stroll through the park

The stroll-through-the-park technique works well for the description of anything that is primarily physical in nature. The idea is to "walk" through the "thing" in some logical pattern, describing what you see along the way.

When describing a room full of equipment, you would start from the front door and "walk" around the room, stopping to describe each piece of equipment as you come to it. If you are physically describing a single piece of equipment, you can use the same approach by starting at the outside cover and "unwrapping" the equipment in layers. Describe what you see as you peel each layer away. By doing this, you are "walking" from the outside to the middle of the equipment, and the audience can easily follow your description.

3.5.4 Time sequence

The time sequence organization works well for the following:

- Description of a process involving a sequential series of activities;
- Coverage of past, present, and future considerations;
- Description of any grouping of events that has already taken place.

Grouping events in a time sequence allows the audience to follow the events logically. Naturally, if there are two or more independent and unrelated series of events that happened to occupy the same general time period (for example, the World Series and a labor negotiation), it would be very confusing to your audience to jump from one series to the other just to keep the whole set of events in absolute time sequence. For this case, it would be best to describe one series of events in time sequence and then describe the other series in its time sequence.

3.5.5 Background to problem to solution

Quite often, the purpose of a technical briefing relates to the resolution of some problem—whether or not the problem is or can be solved. It may be a long-standing problem that your new design is supposed to solve for a customer. It may be a problem encountered during the design process that has to be fixed before the job can be completed. It may be the failure of a piece of delivered equipment. For any kind of "problem resolution" talk, the background-to-problem-to-solution approach works very well.

First, you give the audience the general information that it needs to understand the nature or importance of the problem. This part normally includes any special vocabulary, a description of the way things would be if the problem did not exist, the history of the evolution of the problem, and the way the problem relates to other factors of importance to the audience.

Next, you describe the problem, using the vocabulary and other background information that you have just given the audience. By the end of this part of the talk, the audience should understand enough about the problem to appreciate the nature of the solution you are about to present.

Finally, you deal with solving the problem. If the problem has been solved, describe the nature of the solution. If the problem has not yet been solved, describe the way in which it will be solved. If there is more than one solution and approval is needed to proceed, lay out the alternative approaches and describe the solution you recommend. If the problem cannot be solved, explain why not (and try to prove that it is not your fault that straw cannot be spun into gold).

This organization is powerful in that it first makes the audience smart enough to appreciate the problem, then gives them the bad news about the problem itself, and finally gives them the good news (if any) about how you have helped or will help them get rid of this horrible problem.

3.5.6 Increasing power of arguments

We humans are most happy in a state of constantly increasing expectations. Therefore, when you have a number of things to say on a subject and there is no overriding reason to organize them in any other order, save the best for last. If you are describing mountains, start with molehills and proceed to Mt. Everest in order of increasing height. If you are listing problems, proceed from the most trivial to the thorniest. If you are selling something, list its simplest virtues first and proceed to the strongest advantages over the competition in order of increasing importance to the customer.

3.6 The conclusion

The most important part of the talk is the conclusion. It is the punch line of your talk. It is the part in which you try to cement facts and opinions in the minds of the audience and move your listeners to take the action you want them to take.

The conclusion needs to accomplish three tasks in support of the goals of the talk, listed as follows:

- Review and summarize material;
- Make a call to action;
- Close the talk.

3.6.1 Tell them what you told them

There it is again, the old saw, "Tell them what you are going to tell them, tell them, and then tell them what you told them." In the introduction, you gave the audience a mental "form" with blanks to be filled in with the detailed information supplied later in the body of the speech. The goal then was to be sure that the audience would not become lost in the details of the body.

Now, you will list again your major points to make sure that the audience filled the blanks in the form properly. By reiterating the structure first presented during the introduction and then expanded upon during the body, you will give the audience confidence that it got what it was promised. You also focus attention on the *whole* of the information passed. This makes the details easier to remember because they are remembered in context.

One of the most significant differences between the way machines and humans handle information is the way each relates details to the big picture. Machines remember every little detail and generally deal with the big picture only as a clean, mathematical characterization of the totality of those details. Humans, on the other hand, ignore or quickly forget the little details but have great ability to see and remember subtle trends in the data. Specific details are typically remembered only as they can be related to the big picture.

As part of the material review function, it is an excellent idea to repeat the two or three things you have decided the audience should remember to take the action you want taken.

3.6.2 Sound the call to action

As every good salesperson knows, you have to ask for the order. Remember that the whole reason you are giving a technical briefing is to try to get the members of the audience to take some specific action. If you do not ask them to take that action, they probably will not.

Is this a sales pitch? If so, literally ask for the order.

Is this a talk to convince? If so, tell the audience what it is that they should now be convinced is true.

Is this a talk to give the audience some new skill? If so, tell them what new skill they should now have and ask them to use it properly.

3.6.3 Close the talk

A proper conclusion should give the audience a feeling of closure. One characteristic of an excellent talk is that the audience members feel a sense of satisfaction, even if they do not recognize it at the intellectual level. *That which was promised has been delivered.* The amount of material presented was appropriate to the purpose of the talk.

Some speakers like to close a talk with the same statement or visual aid used to open it. Although this is not always appropriate, it is powerful when properly used.

3.7 Building an outline

The earlier parts of this chapter described the end product of the outlining process. Now, we will discuss the specific steps in the outline preparation process. It must be emphasized that there are many different ways to approach the outlining process. No way is the best way for all. The suggested approach described in this section is, however, a good starting point since many people have found it helpful. In addition, it avoids the writer's block that affects so many new technical briefers. The suggested approach has five parts:

1. Listing of subjects to be covered;
2. Establishment of the framework;
3. Organization of your material;
4. Segmenting of your visuals;
5. Development of your presentation outline.

Remember that all of these steps in the outline development process follow the preliminary talk development steps described in Chapter 2. Before you begin outlining, you should have the answers to the "10 things to do before you start" on paper.

3.7.1 List of subjects

After you have determined the goals for your talk and limited the subject, it is time to list the specific items of information that should be included in the talk. Just put them on a piece of paper as they occur to you; forget about their order at this point.

When you have listed everything that first comes to mind, review the list along with the notes you made in the "before you start" process. This will help you find other items that should be added to the list.

This is a brainstorming process, and nothing should be removed from the list until you are convinced that everything you need to cover is there.

3.7.2 The framework

When you have listed the subjects to be covered, it is time to build a framework for the talk. There are many acceptable techniques for this step, including several very elegant computer programs that allow you to complete the whole process without touching a pencil. The process, however, is basically the same whether done on paper or "on the tube." For ease of understanding of the process, it will be described here as a manually completed exercise with pencil and paper.

Start with a large, blank piece of paper and a pencil with a good eraser. The paper can be any size you like, but it should be large enough to contain all the information you plan to write on it.

Write introduction, body, and conclusion on the page, dividing it into roughly a quarter each for the introduction and conclusion with the balance for the body, as shown in Figure 3.3. This is the basic framework of the talk, but you still need to add a framework for the organization of material within the body.

Pick an organization approach for the body of the talk from those listed in Section 3.5 and add the structure for that approach to your piece of paper. Figures 3.4, 3.5, and 3.6 show how the talk framework might appear for three different types of body organizations.

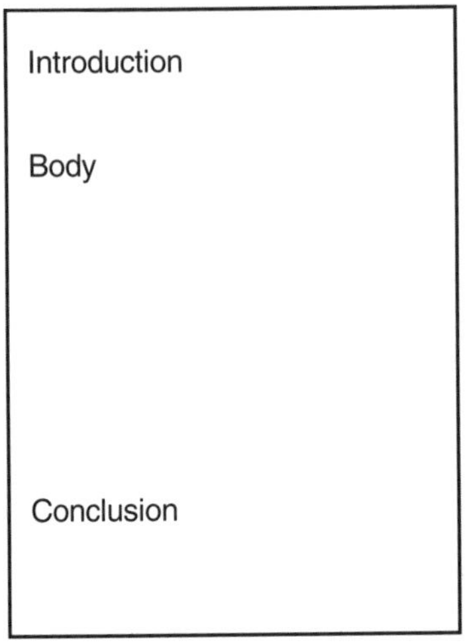

Figure 3.3 Basic talk framework.

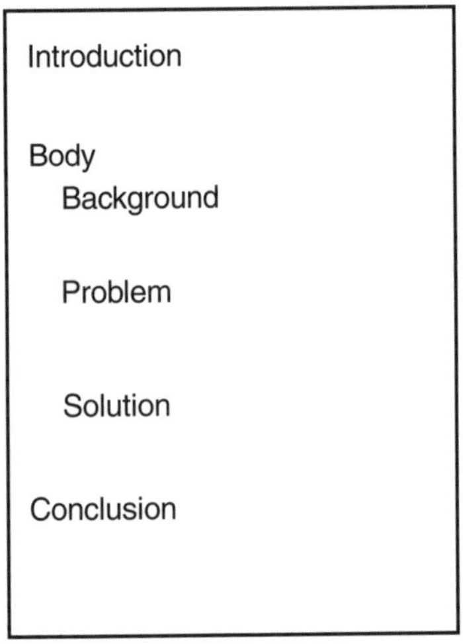

Figure 3.4 Framework for background-to-problem-to-solution organization.

Introduction

Body

Overview

Part 1

Part 2

Part 3

Conclusion

Figure 3.5 Framework for general-to-specific organization.

Introduction

Body

Antenna

RF section

IF section

Audio section

Conclusion

Figure 3.6 Framework for input-to-output organization.

So far you have randomly listed a bunch of information to be covered and established a framework for the organization of this information. Now, the two can be combined.

3.7.3 Organize the material

Rewrite the material from your list onto the framework you have developed, trying to make it fit the format. As you get most of the material in place, you will find that some of the items you originally listed do not really fit and can be thrown out. You will also find that there are some other items that must be added to fill in the whole story.

This is typically a messy and reiterative process making significant use of your eraser. It will, however, allow you to see the flow of your material in the way in which you plan for the audience to see and hear it. When you have finished making such an outline, you will be ready to segment your visual aids and make a presentation outline.

3.7.4 Segmenting your visuals

There *are* rare occasions when visual aids are not appropriate to a technical briefing, but they are so rare that the preparation of visuals is considered a given in the preparation process. Once you have decided what to present and in what sequence, you need to consider how to divide the information into individual visual aids (for example, slides). There will generally be at least one introduction visual, one conclusion visual, and one visual for each major subject covered. Beyond this minimum number, there will be additional visuals required for each circumstance in which a single visual is not adequate to cover the information.

Chapter 4 discusses the process of designing visual aids, including the limiting of information on each visual frame. It also deals with ways to organize the sequence of visuals on a single subject and the appropriateness of various visual media.

3.7.5 The presentation outline

The outline you prepared while organizing your material contains all of the information you plan to present and shows the order in which you plan to present it. This outline may not be appropriate, however, for the paper you want to have in your hand while you are actually *making* the briefing. This paper, which will be your "friend in need" while you are on your feet, deserves separate consideration. Its purpose is not to organize material but to allow you to speak in an organized way.

The presentation outline can take various forms for different types of briefings and briefing situations. Some speakers never use any type of written presentation outline, depending instead on the order of their visual aids to keep their presentations organized. Almost all speakers depend on the order of their visuals to some extent, but some speakers prefer to have a written outline under any circumstances. Appendix B, reprinted from a magazine article entitled "Speech Notes: How and When To Use Them" [1], covers the general field of speaking notes. However, there are special considerations for speaking notes (in effect, a presentation outline) in a technical presentation.

A list of visuals

Since all but the shortest technical briefings typically have a significant number of visual aids, there are logistical considerations that make it highly desirable for the speaker to have a descriptive list of slides on the lectern. The list should include the following for each visual aid:

- An identifying number (which is also marked somewhere on the visual itself);
- The title of the slide;
- A note about the nature of the slide (whether it is a photo or words, for example);
- A brief statement that will help you differentiate that slide from others that are similar.

Such a list will allow the speaker to deal professionally with any of the following circumstances:

- When the actual visuals are handled by someone else (for example, when someone else must flip viewgraph slides because the situation in the presentation room requires the speaker to be physically separated from the overhead projector);

- When someone drops the visuals and gets them out of order five minutes before your briefing is due to start;
- When you are asked a question during the question-and-answer period that can be answered best by use of a single slide located somewhere down in the stack;
- When an unexpected change in circumstances requires that you eliminate some part of your talk (for example, the speaker right before you covers one of the points you thought you had to cover).

Notes for statements

Even though most of the talk may be presented in a way that is tightly related to material on visual aids, it may be appropriate to make a carefully worded introductory or concluding statement. A speaking outline or even (rarely) a text for the statement will allow the speaker to make the proper points in the intended way.

Talks from real-time modified visuals

Often, it is effective to generate visual information while you talk. For example, you might make a blackboard drawing, write on a viewgraph slide with a special pen, or add information to a flip chart while you talk. In these cases, it is an extremely good idea to work out ahead of time what information you will write or draw. Notes showing this information should be part of the speaker's presentation outline.

3.7.6 Developing an outline with a presentation package

It is tempting to just sit down at the computer and start whacking out an outline, start making the slides you know you'll need, or grab another briefing off of the computer and move the slides around. Unfortunately, if you start the process this way (without completing the "10 things to do before you start"), you are likely to find yourself trying to modify a massive amount of information to get it into some semblance of order.

This temptation comes from the ease with which you can move material around in computer programs. Yes, modification is simple. However, it is easy to get lost in the details if you don't start with a well-understood purpose. Therefore, you should either start with a pencil and paper until you know where you want the talk to go or use the computer to make a rough draft. If you are comfortable using the computer (i.e., if you type well enough to join a "chat room" on the Web), open up a word processing program and do your preliminary steps there.

A significant advantage to using a computer at this stage is that you can easily rearrange your outline (when it is new and small) by cutting and pasting. When you get your main points on the screen, you can then go into the draft at any point and fill in more detail. By iterating the process in three or four levels of increasing detail, you are likely to move smoothly from what you want the presentation to accomplish to how the material can best be presented.

Once you have the talk designed, you can start generating the necessary visual aids with some efficiency (and even dignity). Then, you can borrow slides from some other presentation by importing them to the optimum location in the flow of your briefing.

Reference

[1] Adamy, D., "Speech Notes: How and When To Use Them," *The Toastmaster,* Vol. 44, No. 8, 1978, pp. 11–13.

4

Visual Aids—Designing and Using Them

It is difficult to imagine a technical presentation without visual aids because so much of the information passed from the speaker to the audience involves concepts that cannot be handled without visual images. To be effective, however, the visual aids must be selected, designed, and used properly.

This chapter discusses the various kinds of visual aid media that are available and the strengths and weaknesses of each for presenting various types of information, in various types of talks, and in various types of speaking situations. It also includes the mechanics of designing individual visual frames.

For convenience, the term *frame* will be used to denote the material that the audience sees at any single time. Although this term is widely used to mean a single picture in a movie, it can apply as well to a single overhead projector slide, a single flip chart page, or a block of material on a blackboard.

The actual creation of visual aids is not considered in this chapter, as it will be covered in detail in Chapter 5.

4.1 General considerations

Regardless of the visual aid medium chosen, certain rules apply to the use of a visual aid; they are listed as follows:

- The audience must be able to read the visuals.
- The audience should be able to *quickly* grasp the information on each visual aid.
- The rate of presentation of visual aid frames must be appropriate.

4.1.1 Readable visuals

As a rule of thumb, the letters on a visual aid (as presented to the audience) should be one-inch high for each 30 feet to the most distant member of the audience. This size corresponds to that of the block letters on the 20/40 line of a standard eye chart. For direct media (such as a blackboard or flip chart) this is, of course, the height of the letters as drawn. For projected media (including 35-mm slides and overhead and computer projectors), the placement of the screen relative to the projector becomes a part of the equation. Line art should be presented at equivalent size; that is, the thickness of lines in drawings becomes a part of the equation.

In its brochure, *Planning and Producing Slide Programs,* Eastman Kodak points out that the comfortable viewing size for letters must be based on the *effective* size of the letters. As shown in Figure 4.1, the effective letter size for lowercase letters is different from that used for uppercase letters. For lowercase, the effective size of letters excludes the "stems" and "tails." Kodak also adds the rule that letters should be no less than one-fiftieth of the height of the full viewing area of a slide. This means that if the original artwork from which a slide is shot has a total height of one foot, individual letters should be no less than one-quarter-inch high—regardless of the distance to the farthest audience member.

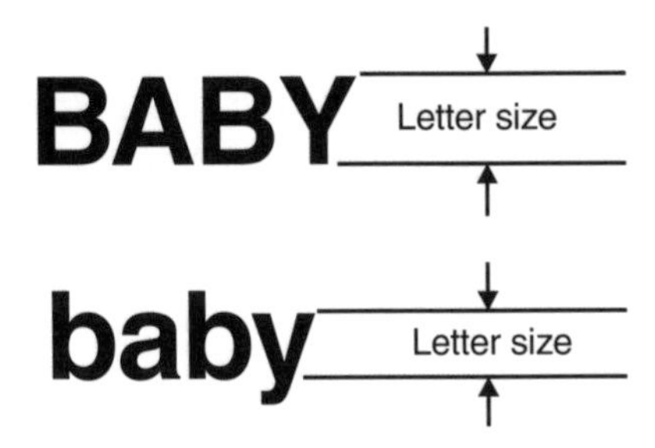

Figure 4.1 Effective letter size.

It should be noted that the standard block letters used on eye charts are very easy to read when compared to letters from more decorative type fonts. For the New York font, slightly more size is required (approximately one and one-quarter inch per 30 feet to the most distant audience member). Old English letters must be two inches high to be read with the same level of comfort as one-inch block letters. These numbers were determined from empirical tests of comparative legibility, using special eye charts generated with computer graphics.

Line art should be presented at equivalent size; that is, the thickness of lines in drawings should be at least the thickness of the lines making up the letters on the visuals. In a series of "eye tests" made while doing research for this book, it was found that one-eighth inch of line width for each 30 feet to the most distant audience member gave approximately the same level of viewing comfort as one-inch-high letters at the same distances.

The size of photographs presented should be such that the detail you want the audience to see is of equivalent size. This, of course, is very subjective; you will need to look at photographic material from different distances to determine if you can easily see what you should see from the picture.

The one inch per 30 feet rule is not absolute. The nature of the material, the level of light in the room, the colors used, or other considerations may require larger, or allow smaller, presentation text. The important thing is for the audience to be able to *comfortably* read the material presented. The only way to be sure the size of text is correct is to check it out personally. Adjust room lighting to what it will be when you make your presentation, put up a typical visual aid, and sit down in the most distant seat in the room to see if you can read the visual comfortably. If you have to squint to see the detail, your audience will not be able to read the visual comfortably, and you should increase the size. It is far better to err on the side of too large a presentation than too small a presentation of the material on visual aids.

A quick and easy test of visual aid readability (that will work in most cases) is to lay page-size paper copies on the floor at your feet. If you can comfortably read them while standing erect, they will probably be comfortably readable for your audience.

The contrast on visual aids should be high enough so that the audience can easily read them. Where photographic processes or projectors are involved, pay close attention to focus. When colors are used, the colors must be chosen for high contrast.

A survey of the managers of several industrial graphic arts departments consistently yielded the following guidelines on the selection of colors for visual aids:

- Use basic colors only (red, orange, yellow, green, blue, or violet), and avoid using variations.
- Use lighter shades of colors for background when black letters or lines are used.
- Use darker shades of colors for background when white letters or lines are used.
- Be careful how you use the color red. It has hostile implications and should not be used to show something you are trying to sell. It is ideal to indicate danger areas.
- Never use more than four colors on a single graphic aid.
- Highlight key words or phrases with yellow or orange.

M. Grumbacher Inc., which makes paints and supplies for artists, has published a "color compass" that is used as a standard reference for color combinations. Figure 4.2 shows the color combinations that provide high contrast, as indicated on this color compass. As with all other artistic

	Red	Orange	Yellow	Green	Blue	Violet
Red		Poor	Good	Best	Good	Poor
Orange	Poor		Poor	Good	Best	Good
Yellow	Good	Poor		Poor	Good	Best
Green	Best	Good	Poor		Poor	Good
Blue	Good	Best	Good	Poor		Poor
Violet	Poor	Good	Best	Good	Poor	

Figure 4.2 High-contrast color combinations: These basic colors are the ones most recommended by graphic designers for visual aid use. The chart shows which colors contrast well (and not so well) with others. Use either the top or side column for reference.

concerns, the acid test is whether or not you can read the visual aids easily from the appropriate distance under the light conditions that will exist in the briefing room.

4.1.2 Limiting the material on a single visual aid frame

The reason that you use visual aids is not just to transfer the information they contain to the audience—that is the purpose of written material. A properly designed visual aid should give the audience members a framework to appreciate what you say while they are looking at your visual aid. This means that your listeners should be able to grasp the significance of a visual aid in a small part of the time you are presenting it to them. Then they have time to listen to what you have to say about the material they have just read and to copy any desired part of the material if they are taking notes.

4.1.3 Rule of six

A guide to the appropriate level of material on a single visual text frame is that there should be no more than six quickly read phrases. If a block diagram is presented, it should have no more than six blocks on a single frame. If a sketch or graph is used, it should be limited in scope so that no more than six items of information are presented to the audience in a single frame. If a photograph is used, its composition should be such that the eye quickly focuses on the information the speaker wants the audience to learn from the picture.

4.1.4 Nested visuals

When complex information must be presented, there is a natural temptation to include all of the required detail on a single visual frame. This minimizes the work for the presenter but makes the presentation impossible to follow by the audience. A much better way to present complex information is by using nested visuals.

The nested visuals concept starts with the presentation of a single overview frame with the "big picture." This is a simplified diagram that divides the whole device or concept into a few (fewer than six) major parts. Then, a separate visual frame shows each of those major parts to the next level of detail, a few major subparts. If necessary, additional frames are used to bring in finer and finer detail.

By returning to the big picture before presenting each of the major parts, the speaker keeps the audience from getting lost in the detail. By the same token, the speaker should return to the major part level before describing each of the major subparts.

Although this process involves significant preparation by the presenter, it pays great dividends in presentation effectiveness for two reasons. First, the audience can read the slides. Second, the audience will be forced to consider the material in the order in which the speaker wants it considered. A significant side benefit is that the presenter has to think hard about how the audience will "get into" the information, and this communication approach will show throughout the entire talk.

An example of nested visuals is shown in Figure 4.3. The "system" shown in Figure 4.3(a) could be any kind of hardware or process too complex to be shown in a single visual frame. Figure 4.3(b) shows a portion of

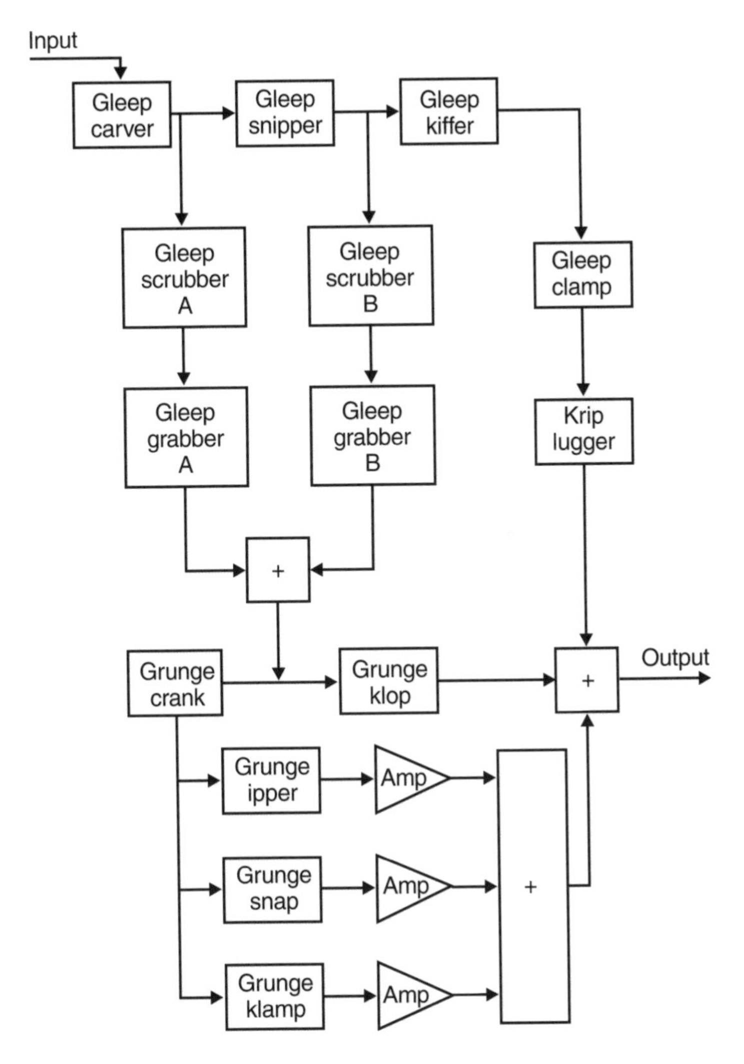

Figure 4.3(a) A diagram too complex for one visual frame.

the set of properly nested visual aids with which the system could be described. The partial set shown covers only the "gleep subsystem." The full set of visuals would include the same level of detail for each of the major parts shown in the overall concept slide.

Properly used, the three visuals shown in Figure 4.3(b) would be shown in the indicated order. Then, the first visual (the overall concept slide) would be presented a second time, followed by the nested visuals

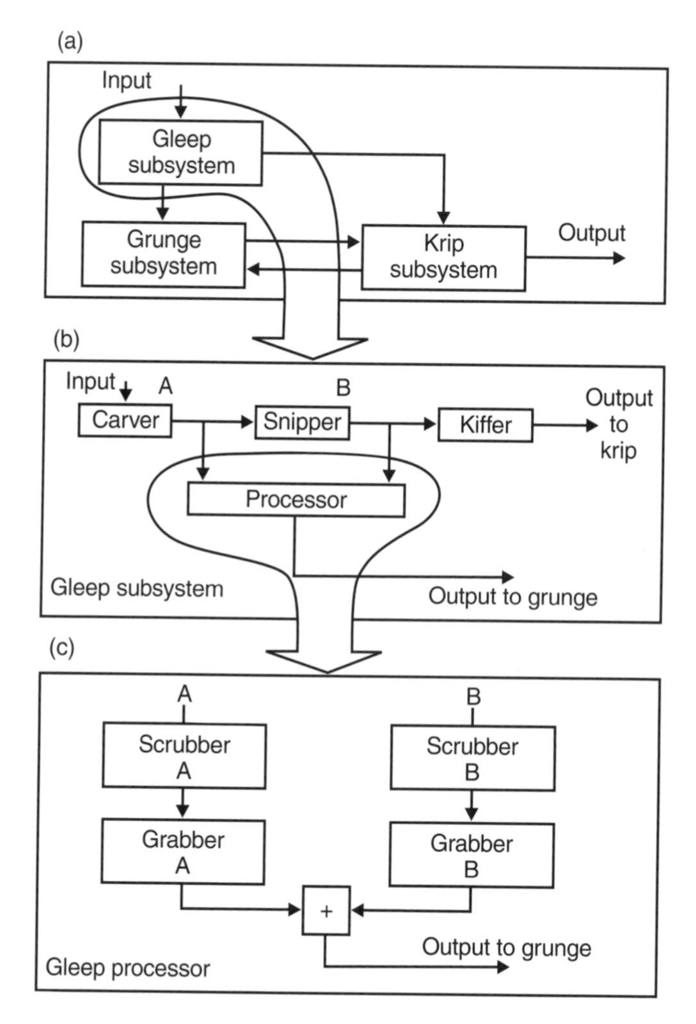

Figure 4.3(b) Nested visuals to explain: (a) overall slide concept; (b) first-level detail slide; and (c) second-level detail slide.

for the "grunge subsystem." Then the first visual would be presented a third time, followed by the nested visuals for the "krip subsystem."

4.1.5 Rates of presentation for visual aids

The rate at which visual aids are presented is very much related to the visual aid medium used. The rate must be comfortable for the audience, with a new visual frame occurring before the audience gets tired of the old one—but not so soon that the audience is unable to absorb the material. In the most effective presentations, the rate of presentation of visuals is varied over the course of the talk but does not exceed a two-to-one range. Some variation in pace helps keep the audience interested, but too great a variation breaks the rhythm at which the audience expects to receive new inputs.

4.2 Visual aid media

There are many types of visual aids available, and choosing one is an important decision that the speaker must make early in the preparation process. The decision is always limited by cost, availability of equipment, and availability of preparation time. It should not be limited, however, because of lack of imagination of the speaker or by the speaker's lack of knowledge about what is available.

The following visual aid media will be covered in this chapter.

- Direct computer projection;
- Overhead projector slides;
- 35-mm slides;
- Moving media (movies and videotapes);
- Flip charts;
- Blackboard/whiteboard;
- Handout material;
- Physical items to be passed around;
- Three-dimensional models;
- Demonstrations.

This is not a complete list of all of the visual aid media available, but it is sufficient for our purpose, which is to help you choose visual aid media properly and use them effectively.

4.2.1 Comparison of media

Table 4.1 shows the selection criteria for all of the visual aid media discussed in this section. It must be emphasized again that the considerations presented are not absolutes; they are just common practice guidelines.

4.2.2 Special techniques and considerations

There are tricks of the trade that are used by experienced technical briefers to make their visual aids particularly interesting or to handle unique information requirements. Appropriate techniques and considerations are included in the section describing each type of visual medium. Although techniques are only described in the context of the single medium in which they are most often used, it will be obvious that most of these techniques can be adapted to other media.

4.3 Direct computer projection

Direct projection of presentation visual aids from computer files has become increasingly common over the last few years. The presentation is developed in one of the presentation formats (PowerPoint, for example). Then, a special projector is connected to the "VGA" port on the computer. Figure 4.4 shows two typical commercially available direct computer projectors. The unit in Figure 4.4(a) is mounted to the ceiling of a conference room for maximum convenience in a permanent use setting. These units can also be brought into a temporary setting and placed on the conference table. The projector in Figure 4.4(b) is one of the new lightweight portable units that weigh about five pounds.

The direct computer projector mimics the computer screen, projecting it onto a screen. The presenter can move through the presentation by commands from the computer keyboard or mouse or by use of a remote control device compatible with the projector.

These projectors are available in various resolution levels. They are also available over a range of projection power to make them appropriate for different audience sizes. There is continuing competition among suppliers to reduce the size of the lower power units to make them more convenient to carry to briefings.

Table 4.1

Visual Aid Medium Selection Guidelines

	Conditions Affecting Medium Selection					
Visual medium	**Audience size**	**Room light**	**Timeliness of material**	**Visual aid equipment available**	**Nature of material**	**Other**
Direct computer projection	5–1,000+	Can be dimmed	Must be current	Computer, computer projector, and screen	Photographs, words, and line drawings	
Overhead projector slides	5–100	Can be dimmed	Must be current	Projector and screen	Mainly words and line drawings	Overlays appropriate to the subject
35-mm slides	25–2,000+	Can be dimmed	Not critical	Projector and screen	Photographs, words, and line drawings	
Moving media	5–1,000+	Can be dimmed	Not critical	Projector and screen	Graphic and important action	
Flip charts	Less than 15	Cannot be dimmed	Must be current	Flip chart stand	Mainly words and line drawings	
Blackboard or whiteboard	5–50	Cannot be dimmed	Must be extremely current	Blackboard or whiteboard	Words and very simple line drawings	Classroom situation
Handout material	Any size	Can be left bright	Must be current	None	Easily reproducible	Informal briefing or as backup data
Pass around items	Less than 20	Can be left bright	Must be current	None	3-D appearance is important to talk	Must be nonbreakable or expendable
Three-dimensional models	Up to 30	Can be left bright	Not critical	None	Physical interrelationships are important	
Demonstrations	Up to 30		Not critical	None	Action is important to subject	Should have plenty of time

Figure 4.4(a) Ceiling mounted direct computer projector.

The earliest computer projection approaches involved a flat plate that acted as a computer-driven overhead projector slide. In general, these devices provide much less brightness than the later direct projectors. Accordingly, they have rarely been used since the direct projectors were introduced. Chapter 6 focuses on techniques for use of these projectors and the software available to prepare material for presentations on them.

4.3.1 Advantages and disadvantages

There are many advantages to the use of direct computer projection:

Figure 4.4(b) Portable computer projector.

- It is easy to add color to a presentation. Since almost all modern computers have color screens, a wide range of colors is available for use in any way the presenter sees fit. You can use color text, colored line drawings, color-coded charts, or even scanned-in color photographs.
- Software is available to dramatically move from one slide to another—fading or cutting across the old slide as the new slide appears.
- Slides can be made directly from other computer files (including graphics, spreadsheets, and tables).
- Any material available in hard copy can be scanned and added to any slide.
- Video clips can be projected from the same projector and can even be added as small insets to slides containing other information.
- A complete briefing can be carried in a few floppy disks or hard disks (Zip disks, for example) avoiding the need to transport bulky packages of overhead projector slides.
- You can project real-time images of computer processes in action. For example, in one particularly effective briefing on video

compression, a video clip was subjected to various types and levels of compression and presented to the audience in real time.

There are, however, some disadvantages to direct computer projection:

- There is no universal standard interface, so it is possible to arrive at some distant briefing location to discover that your briefing is in the wrong software or that your computer does not talk to the projector available at that location. This makes it an excellent idea to take along a set of backup slides in some more universally available medium (e.g., overhead projector slides).
- Direct projection systems are far from universally available, particularly when you are traveling internationally. This means that you must take along your own computer and projector—and don't forget the power adapter—if you want to depend on this technique.
- Graphic data, particularly color graphics, requires a great deal of memory. Thus, it is very often impractical to store presentations on floppy disks. The problem is that there is no stable standard for higher density storage media (i.e., with more capacity than a 3.5-inch floppy disk). Every time one becomes widely available, it seems to be supplanted by a new, higher density storage medium.

4.3.2 Selection criteria

Direct computer projection may be the best visual aid medium choice when one or more of the following conditions apply:

- Your material is developed using presentation software.
- A computer and compatible projector will be available.
- A projection screen will be available.
- Color is important to the presentation.
- The material presented must be current.
- Real-time computer screen data can be included in the presentation.

- Lights in the room can be dimmed.
- The audience is of moderate to large size (5–1,000+).

4.3.3 Rate of presentation

Because slides can be changed quickly with direct computer projection, there is a great deal of flexibility in presentation speed. The audience must still be able to read a slide in a small part of the time it is up, but it is practical to "build" the information over several slides. Each sequential slide will have additional information—for example, the first slide may have one bullet, and the second slide will include the first bullet but add (highlighted) the second bullet. With this technique, each slide may remain up only about 10 seconds, but the time to present a full page of information (i.e., six bullets) to the audience should still be in the range of one to two minutes.

4.3.4 Special techniques and considerations

Most remote control units for direct computer projectors include provision for an arrow that can be moved around the screen as a pointer. This allows the presenter to be almost anywhere in the room, as long as there is line of sight between the remote control unit and the projector. Depending on the pointer control mechanism, it may not be easy to quickly and smoothly move the arrow to the exact desired point. Thus, it is important to practice enabling, moving, and hiding the pointer arrow before the presentation. Also, be sure to determine where the remote control unit must be located. Consider, for example, that many will not work if you hold them below the level of a conference room table. Depending on the situation and the audience, you may want to use a laser pointer or a regular physical pointer rather than the arrow on the screen.

There is usually a dedicated computer for a conference room with a direct computer projector, so you can take your presentation to the room on a floppy disk, Zip disk, or similar memory device. Many organizations have their internal computers connected through a local area network, so it may be practical to develop your presentation at your own desk (or some available work area) and store it in a common memory location from which it can be downloaded by the conference room computer.

When you are giving a presentation on the road and cannot depend on the availability of a computer or projector, it is normal practice to have the presentation loaded onto a laptop computer and to bring along an appropriate projector. For large audiences, a high-power (also heavy) projector is required, but for smaller audiences (less than 20) one of the new lightweight projectors may be the best solution. The size, weight, and cost of portable projectors are decreasing almost monthly.

Handheld computers (smaller than a laptop) can output a presentation to a direct projector but may not allow you to create or edit your presentation. Sometimes the first slide can be modified to allow you to change the title slide for each audience.

As with all projected visual aid media, be aware of your shadow on the screen and of the potential to project the visual aid onto your face or body. In general, it is best to stay out of the area between the projector and the screen as much as possible. Reaching in from the side to use a pointer is fine, but standing with the technical material gloriously spread across your face is distracting to the audience.

4.4 Overhead projector slides

Despite the rapid rise of direct computer projection, the most commonly used visual aid medium is still the good old overhead projector (or Viewgraph). It projects page-size transparent slides onto a screen. Figure 4.5 shows a common type of overhead projector. There are many different configurations of overhead projectors on the market, from the very large "long-throw" projector that can fill a large screen to the lightweight portable unit that will fit into a large briefcase.

4.4.1 Advantages and disadvantages

The overhead projector is available in most technical briefing situations, can be used in a moderately lighted room, and uses slides that can be quickly and easily made on copy machines. It can use also professionally developed slides including photographs or line art. Overhead projectors are convenient to use because you can face the audience while using them, yet see the same thing on the projector slide that the audience sees on the

Figure 4.5 Overhead projector.

screen behind you. You can also use overlays or special pens to add information to a visual frame while you are talking about it.

Nevertheless, the overhead projector has several disadvantages as a visual aid medium. Large numbers of slides are bulky to carry and the motion required to change from one slide to another is time-consuming and distracting to the audience. Moreover, professionally made overhead slides are quite expensive, particularly when they are color photographic positives.

4.4.2 Opaque projector

In earlier years, the opaque projector, a device related to the overhead, was popular. The opaque projector would project the image of a piece of opaque material (such as a piece of paper, photograph, or page of a book) onto a screen. These devices were large and heavy and, because they used very bright lights that generated lots of heat, required a loud fan for cooling. They would also occasionally enliven a classroom situation by setting fire to the piece of paper they were projecting. They almost disappeared when it became practical to make transparent overhead projector slides on copy machines.

4.4.3 Selection criteria

The overhead projector may be the best visual aid medium choice when one or more of the following conditions apply:

- A projector and screen will be available.
- The material presented must be current.
- Overlays are appropriate to the subject.
- The lights in the room can be dimmed.
- The audience is of moderate size (5–100).
- The material is mainly words and line drawings.
- You cannot be confident of the availability or compatibility of a direct computer projector.

4.4.4 Rate of presentation

For overhead projector slides, a generally followed rule is one slide per minute, although many effective speakers use average slide change rates as low as half or as high as twice that. The limiting factor is the distraction of the audience when one slide is manually removed from the projector and another is put in its place. If information on the slide is sequentially disclosed (using heavy paper to hide part of the material) or overlays are used, it is appropriate to spend more time on a single slide. If someone other than the speaker is changing slides—and that person is very good at the job—slide change rates up to two per minute are effective.

4.4.5 Special techniques and considerations

The techniques described here are overlays, pen on slide, and half-frame slides. The considerations are mounted versus unmounted slides, slide numbering, notes on the mount, and slide orientation.

Overlays

Overlays are additional slides that are attached to the edges of an overhead projector slide so that they can be folded into place to add information to the slide at the appropriate time (see Figure 4.6). The power of overlays is

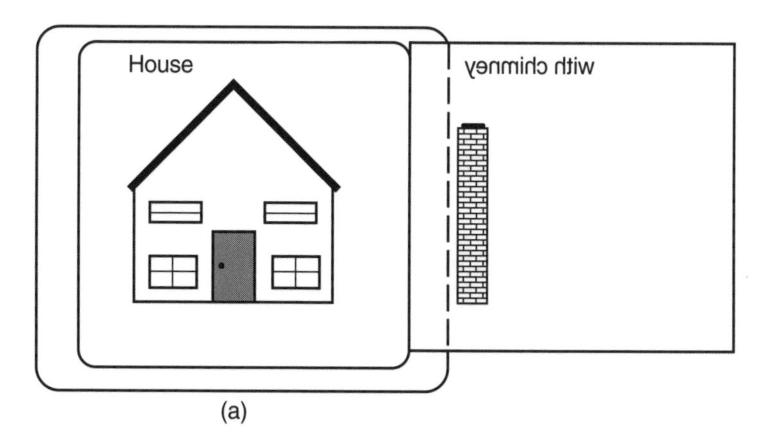

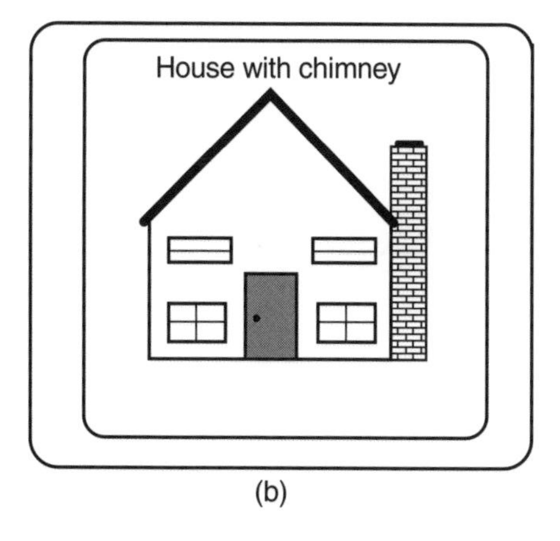

Figure 4.6 Overhead projector slide overlay: (a) basic slide and (b) basic slide with overlay in place.

that they allow the addition of in-context information to a slide format already being considered by the audience. They are very effective for displaying trends in data or the building of a concept, but they can be used to display almost any other type of information. The overlay information is added without the distraction of removing the basic slide, so the audience intuits that the new information is related to the old. Adding information with a second slide often requires the explanation that the audience is seeing the original data with some new data added.

Multiple overlays can be used on the same basic slide. Sequential overlays can be placed on top of each other to build a full picture gradually as shown in Figure 4.7. It is also acceptable to have up to three independent overlays as shown in Figure 4.8. Each one is removed before the next

Figure 4.7 Sequential overlays.

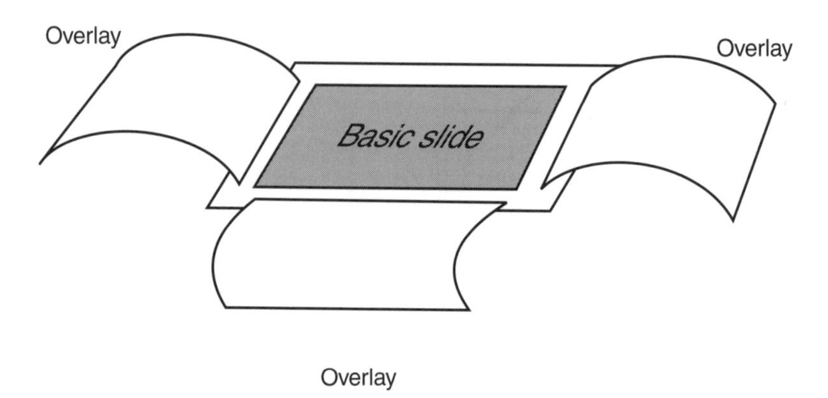

Figure 4.8 Independent overlays.

is folded in place to show alternate configurations or data. The limitation of three independent overlays is dictated by the three available edges on the basic slide. The top edge is not available because the overlay will hit the lens support post and cannot be folded out of the way.

When preparing overlays, be sure that they are in exactly the same scale as the basic slide. The basic slide should be mounted in a slide mount, sometimes called a slide frame. The overlays are not mounted. Each overlay is placed in the proper location over the basic slide, and one edge of the overlay is taped to the basic slide's mounting (free of the corners of the overlay).

Pen on slide

Special pens that write on plastic are available from any stationery store. The pens come in many colors, in several line thicknesses, and with permanent or water-soluble ink.

The technique involves the preparation of overhead slides that do not contain all of the information to be presented, so that the briefer can add information to the slides during the presentation. If water-soluble ink is used, the slides can be cleaned after the briefing and used again.

Pen on slide is used effectively for filling in the blanks on a form as the audience learns the proper procedure. The form can be either an actual form that will be used by the audience elsewhere, or a special form generated by the briefer just to help the audience understand a process. Figure 4.9 shows an example.

Another application involves the preparation of a slide with the problem shown in graphic form and with the graphic answer added by pen as the problem is solved. See Figure 4.10.

For this technique to work well, you need to have the information you will be adding to the slide well worked out in advance. Your markings on the slides may seem spontaneous to the audience, but unless you are blessed with an amazing grasp of the material and coolness under fire, you don't want to take a chance on getting the information wrong. The greatest minds sometimes go blank before an audience. An excellent technique is to have a hard copy of your presentation on a table beside the projector. Add the information you are going to write onto the slides to your copy in clear, large print in a contrasting color.

Link equation work sheet

ERP (dBm)	+30
Xmit antenna gain	+ 8
Space loss	- 98
Atmospheric loss	- 2
Receiving antenna gain	+ 3
Receiver sensitivity	- 65
Margin	+ 6 dB

Figure 4.9 Pen-on-slide "form" example.

Half-frame slides

There is no law that requires every overhead projector slide to be full-page size. If you are making your own slides on a copy machine, you can make them any size or shape you desire. By making each slide fit onto half of an 8.5- × 11-inch page, you can make two slides on the page and cut the page in half. There are more reasons to do this than just to save material, although that is not a bad reason, because the material is quite expensive.

Another reason is that you can split the information on a single screen. You can have a drawing or photograph on the top of the screen while you display a related word slide on the bottom of the screen as in Figure 4.11. This technique is also very powerful when you are using nested visuals. Keep the "big picture" slide on the top of the screen while you are showing each of the detail slides on the bottom. Your imagination is your only

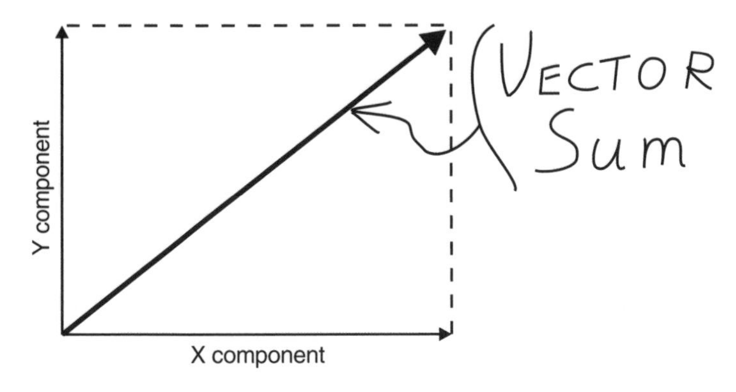

Figure 4.10 Pen-on-slide graphic solution example.

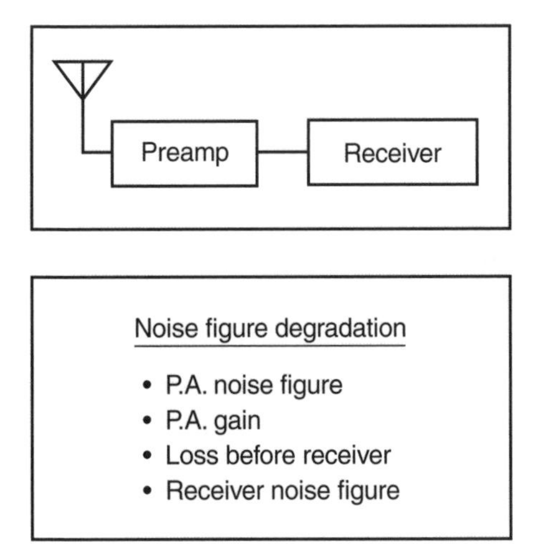

Figure 4.11 Example of split-screen visual.

limitation with this technique—you can even cut up an overhead slide like a jigsaw puzzle!

Mounted versus unmounted slides

You do not have to mount overhead projector slides; however, there are advantages to doing so. The frames keep the slides flat on the screen, limit the projected image area, and provide a convenient place to number the slides. They also can make it easier to place the slides straight on the projector and keep the slides from blowing away if there is wind. (Often the fan in a projector will be the source of significant wind across the projection surface.) A final advantage is that mounted slides are harder to scratch during handling and transportation.

The main reason people do not mount overhead slides is that the slide mounts are very bulky. One hundred unmounted slides can be carried in the same amount of briefcase room as 12 mounted slides. Overhead slide material comes in several thicknesses. If the slides are to be used unmounted, the thickest available material should be specified. If the slides are less than full-page size, they will ordinarily need to be left unmounted.

Slide numbering

Always number your slides. If they are mounted, place the number on the mount. If they are unmounted, write a number in the corner of each slide with one of the special pens described in the pen-on-slide section.

Notes on the mount

Many experienced briefers like to make notes around the sides of mounted overhead slides to help them remember pertinent information to present verbally while they show the slide. This is particularly important if hard-to-remember numerical data is to be added to a slide during the presentation.

Slide orientation

Everyone has been in a briefing in which the briefer placed an overhead slide on the projector upside down then got it reversed while trying to correct the original error. This is particularly problematic with unmounted slides and it is very disruptive to the flow of your briefing.

The best way to assure that you will place the slides correctly is to make some sort of consistent mark in the same relative position on each slide. If the slide number is always placed in the same position (for example, the upper right corner of the slide or slide frame) as it is to be placed on the projector, this will serve the purpose very well. Another common technique is the use of a distinctive title area at the top of each slide.

4.5 35-mm slides

The most widely used medium for large groups or very formal presentations is 35-mm slides. This medium projects film positives mounted in standardized two-by-two-inch slide mounts. Figure 4.12 shows a common type of 35-mm slide projector used for technical briefings. It holds up to 150 slides in a single slide tray or carousel and has a remote controller to advance slides and adjust the focus of the projector. A very large

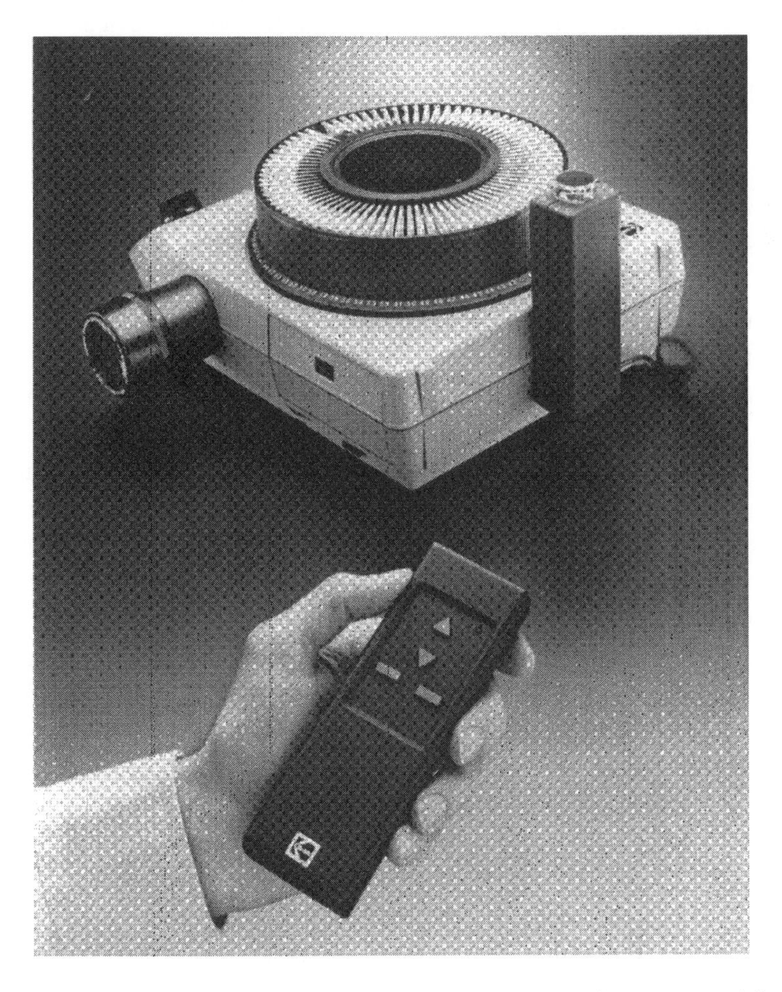

Figure 4.12 35-mm slide projector. (Reprinted courtesy of Eastman Kodak Company.)

number and variety of 35-mm slide projectors are available. They accept various sizes and configurations of slide holders, provide different projection ranges, and trade off operating features against a wide price range.

4.5.1 Advantages and disadvantages

Thirty-five-millimeter slides are small, so it is easy to carry a large number to a briefing. Slide carousels hold large numbers of slides in the proper order, thus simplifying the logistics in the briefing room. The speaker can change them from the lectern by using a remote control device. The time to change a slide is less than that required to change visual frames in almost any other medium. Projectors for 35-mm slides are very widely available, and many are small enough to be carried easily to the briefing site. Slides can be used in portable, automatic slide/tape devices in which the slides are synchronized with an audiotape for a complete canned briefing. There are also unique presentation approaches that only work with 35-mm slides.

The primary shortcoming of 35-mm slides is that they require formally prepared artwork, which is then shot with a camera and the film processed into slides. The logistics, cost, and time delay in generating a set of 35-mm presentation slides make them inappropriate for informal briefings to small groups of people. This problem is being uniquely addressed by several, new computer-based systems that create finished slides directly from computer generated graphics. There are companies in most areas that will produce 35-mm slides from your computer files. They are a little expensive but generally offer fast turnaround.

Another problem with 35-mm slides is that they generally require that the room be darkened to a greater degree than that required for the use of overhead projector slides. Computer-generated graphics should reduce this effect somewhat, since the greater contrast levels associated with the composition of overhead projector slides will be achievable on 35-mm slides as well.

4.5.2 Related media

Two types of visual aid media are closely related to 35-mm slides. One is film strips—basically 35-mm slides with the continuous film roll intact.

The second is lantern slides, which are larger slides with a two-by-two-inch actual viewing area.

Film strips were widely used a few years ago, particularly in schools, and there are some very convenient, small, self-contained display devices in which film strips are automatically synchronized with audiotape cassettes for canned briefings. The main disadvantage of the film strip compared to individual 35-mm slides is that the order of the visuals cannot be readily changed to customize the briefing to different audiences or situations.

Lantern slides have the same advantages and disadvantages as 35-mm slides except that they project a larger visual area, which is sometimes desirable, and that they require projectors that are in less common use than are 35-mm projectors.

4.5.3 Selection criteria

Thirty-five-millimeter slides may be the best visual aid medium choice when one or more of the following conditions exist:

- A projector and screen are available.
- It is not critical that the material presented be very current.
- The lights in the room can be dimmed.
- The audience is of moderate to large size (25–1,000+).
- The material includes photographs as well as words and line drawings.

4.5.4 Rate of presentation

A wide range of frame change rates can be managed when using 35-mm slides, particularly when the speaker uses a remote control slide changer. Although it is quite acceptable to use the one-slide-per-minute guideline, the reduced time for changing 35-mm slides makes higher rates effective as long as the material presented does not change too quickly for the audience to follow. "Dwell times" as low as 5–10 seconds per slide are sometimes acceptable. Section 4.5.5 describes a technique example in which high rates for 35-mm slides are used to great effect.

4.5.5 Techniques and considerations

This section discusses techniques involving sequential disclosure slides, negative image slides, and multiple projectors.

Sequential disclosure slides

The ease and speed with which 35-mm slides can be changed allow the use of a special technique that sequentially discloses information to the audience by use of several slides instead of a single slide. Each slide reproduces the material on the previous slide but adds more. The new information on each slide is highlighted by use of a more vivid color than that used for the old material on the previous slide.

Figure 4.13 shows a simple diagram that would be expected to occupy a single slide. Figure 4.14 (a–e) shows the way that the information can be disclosed sequentially to the audience using five related slides. The new information on each slide can be shown in boldface as presented in Figure 4.14 or in a contrasting color. The new information is discussed as each slide is presented.

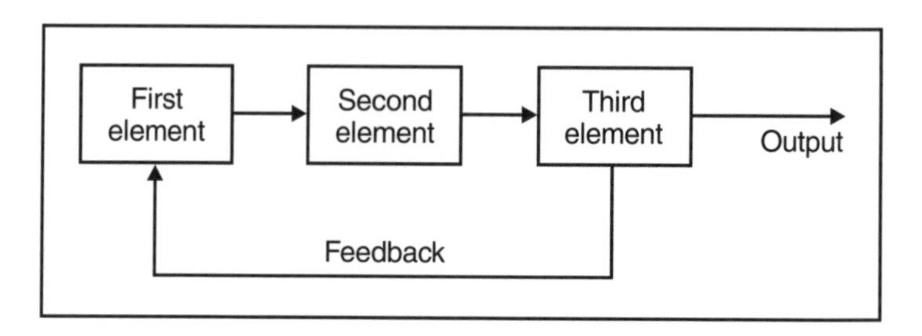

Figure 4.13 A typical slide.

Figure 4.14(a) The first slide in the sequence.

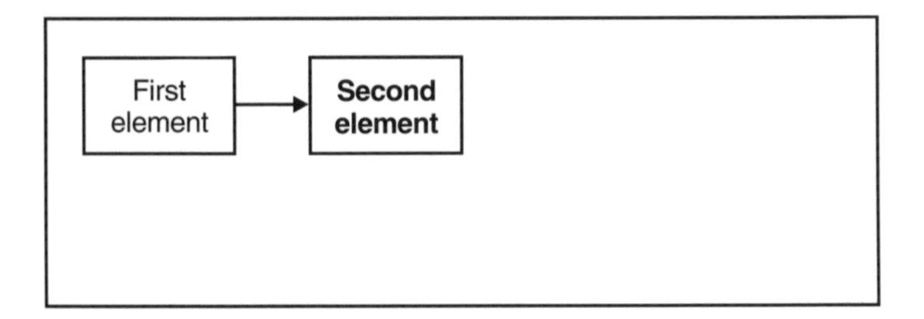

Figure 4.14(b) The second slide in the sequence.

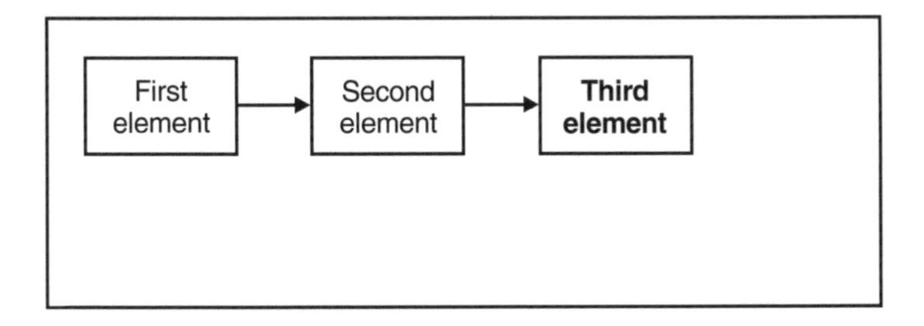

Figure 4.14(c) The third slide in the sequence.

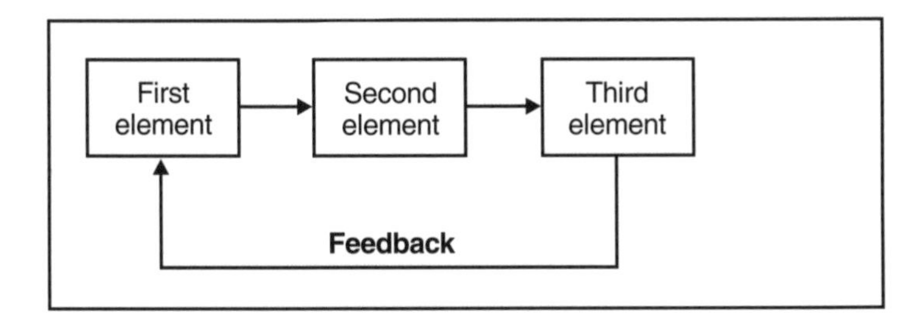

Figure 4.14(d) The fourth slide in the sequence.

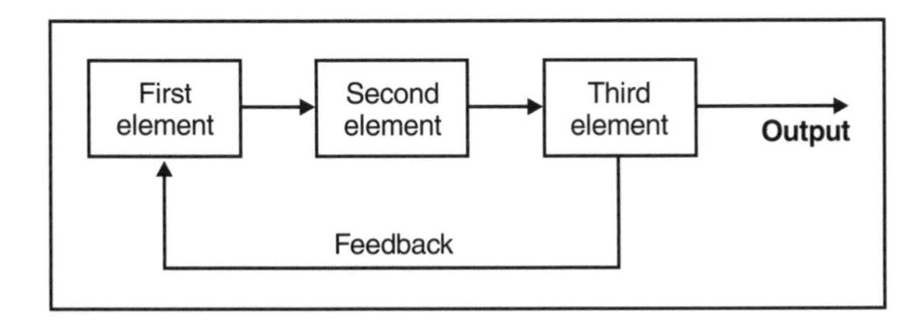

Figure 4.14(e) The fifth slide in the sequence.

The advantage of this technique is that it keeps the audience's attention exactly where the briefer wants it to be more effectively than indicating information on the slide with a pointer. It also eliminates the need to use a pointer, an advantage when the room is dark or the screen is very large.

Negative image slides

One of the shortcomings of 35-mm slides is that the room needs to be dark, making eye contact with the audience and note-taking by the audience difficult. It is practical to keep the light in the room at a significantly higher level if negative image slides are used, since the bright letters and lines against a dark background provide better contrast than normal positive images, in which most of the background is lighter than the information presented. As you will see in Chapter 5, negative image slides are easy to make and particularly compatible with sequential disclosure techniques.

Multiple projectors

Coordinated slide shows using multiple projectors can use either one screen or multiple screens. The single screen approach uses two projectors and a mixer unit that will automatically fade out one slide while fading in a second. The result is a smooth, professional-looking transition between slides. The distraction of the darkened screen between slides is eliminated, making the briefing flow much more comfortably.

The other approach is the multiple screen "extravaganza" that uses from two to nine or more screens. The slides on all of the screens are time-coordinated (either manually or automatically) to give an integrated briefing. All of the screens can show a single wide-angle view of the same scene and sometimes show individual but related information.

4.6 Moving media

Moving media are those that can store and reproduce action. This section focuses primarily on movies and videotapes but also covers direct computer displays. Figure 4.15 shows three typical moving media devices.

Figure 4.15(a) Large-screen TV. (Photo reprinted courtesy of the Sony Corporation.)

Figure 4.15(b) Movie projector. (Photo reprinted courtesy of the Eastman Kodak Company.)

Figure 4.15(c) Large-screen computer monitor. (Photo reprinted courtesy of Apple Computer, Inc.)

4.6.1 Advantages and disadvantages

Everyone loves movies! Audiences typically react with higher interest to the continuous action in moving media. A videotape of someone speaking has an advantage over the real person talking because the listener is intimidated by the steady flow of the talk. There is no option to interrupt the speaker; the speaker will just go on without you.

A well-prepared movie or videotape, with background music and graphic portrayal of events, is an extremely effective way to present information. When much of the message is in the action of what is going on (for example, the performance of an aircraft), there is simply no better way to convey messages.

Even with the simplifications in preparation that came with moderate-cost video recorders, the logistics in preparing a moving media presentation are still much more severe than for other types of visual media. You must plan the presentation minute by minute, direct the production, and edit the film or tape. Typically, the investment of time and money restricts the use of moving media to information that will remain current for a long time, and it is appropriate for presentations that remain unchanged to a large number of groups.

Something to consider is the use of short, unedited pieces of film or tape to cover part of the subject matter in the middle of a technical briefing that is primarily built on a different type of visual. Use the short film clip to illustrate a point. This takes careful preparation and rehearsal to work effectively.

4.6.2 Selection criteria

A movie or videotape is the best visual aid medium choice when one or more of the following conditions are present:

- Projection equipment will be available.
- The material presented is not perishable.
- The lights in the room can be dimmed.
- The audience is of moderate to large size (5–1,000+).
- The material is mainly graphic and includes important action.

4.6.3 Rate of presentation

For moving media, the individual frames are presented quickly enough to seem continuous to the human eye (24 or more frames per second). However, there is a second time consideration, that is the rate at which the information displayed changes. For example, a movie of a flip chart page must remain on a single scene long enough for the flip chart to be read.

4.6.4 Techniques and considerations

The techniques considered here include videotape presentation, film to tape, tape to film, direct display on computer terminal, and mixture of moving and still visual media.

Videotape presentation

Videotapes provide a relatively inexpensive and flexible way to bring moving media to the briefing room. When the audience is small enough that all can comfortably view a normal television screen, it is practical to use a VCR and video monitor to present tapes. This is limited to about

5–10 people, depending on the room layout and the size of the screen on the television set being used.

It is still possible to show videotapes to larger groups. You can rent or borrow a large-screen television/VCR like those used in sports bars. These devices have approximately 4-ft screens that work well for groups of up to 50 or more, and two or more can be slaved together to provide more room coverage. You can also play videotapes with a direct computer projector that also accepts signals from a VCR. For very large groups, you may need to convert the videotaped material to film or rent one of the long throw projectors as shown in Figure 4.16. Again, this depends on the quality of the equipment, the layout of the room, and comfortable ambient light levels.

The international briefers face a particular challenge with videotapes, since there are different formats for different parts of the world. There are videotape players that will accept any of the tape formats, but even these can have problems with the frequency of the power used to run them. For example, a U.S. VHS format tape was played on a special video machine in Sweden. The picture was viewable, but ran at five-sixths of the proper rate (because of the 50-Hz power in Sweden versus 60 Hz in the

Figure 4.16 Video projector. (Photo reprinted courtesy of the Sony Corporation.)

United States). You can have tapes commercially converted to the proper standard for any country, and it is desirable to have this done far enough ahead to allow time to send the tapes to a local contact for a test run while there is still time to fix problems.

Film to tape and tape to film

Material that is on 8- or 16-mm movie film can be converted readily to videotape. Likewise, videotaped material can be converted to film. Both processes are done commercially at moderate prices by large film-processing laboratories.

Direct display on computer terminal

For very small groups (three or less), it is practical to present a whole briefing on a computer terminal. The visual aids are computer-generated and stored in memory. Software that will convert the individual visual frames into an integrated show is available for most computers. Some programs provide automatic sequencing with fade-in and fade-out features. Some even have cartoon characters and stages with curtains that open and close. Some provide a synthesized voice for the briefing, while others require the use of an audiotape recorder with cues for a viewer to advance the slides with a keystroke or mouse commands.

Mixture of moving and fixed visual media

When integrating moving media into a briefing that also uses slides or other fixed media, it is important to coordinate both the information in the two media and the way information is presented to the audience. Typically, the moving medium displays the exciting parts with lots of action while the fixed medium presents the background facts and figures. Any time media are mixed, both must be appropriate to the same audience size and room layout. It is acceptable (but not desirable) to change the light level in the room when switching from one medium to the other. It is much better to choose completely compatible media to allow the smoothest overall presentation.

One particularly powerful media combination is 35-mm slides with 16-mm movie film. This allows all of the information to be presented on a single screen. The briefer starts and stops the movie and changes slides with remote control devices, inserting opaque slides into the slide sequence to darken the screen when showing movie portions.

As mentioned earlier, it is practical to combine moving and still information in a computer-generated presentation. This, in effect, makes a multimedia presentation without all of the logistics of coordinating the operation of different types of projectors. However, all of the input information must be in computer-compatible form, and considerable effort may be required to format the whole briefing to create a pleasing tempo with smooth transitions.

4.7 Flip charts

Flip charts are drawings on large pads of paper mounted to portable easels, as shown in Figure 4.17. Grease pencils or wide felt-tip pens are used to draw figures or text. As the briefer finishes with each visual, he or she flips the page over the top of the easel to expose the next visual.

Figure 4.17 Flip chart visual aid.

Sometimes, nothing but a flip chart will work because of the situation in the briefing room (for example, when no other visual aids are available to the briefer). Other times, it is just the most professional way to handle a briefing to small groups.

4.7.1 Advantages and disadvantages

Flip charts have the great charm of being usable in any kind of room as long as there is enough light to see them. You can either prepare the flip charts ahead of time—or draw them as you are talking. It is also very convenient to add material to a flip chart as you speak. Flip charts provide a permanent record of the information generated in the process of the briefing. Often, when the purpose of a briefing is to generate input from the audience—for example, during a strategic planning session—the flip chart is by far the most appropriate visual aid medium.

The shortcomings of flip charts are that they are ungainly to change (pages must be flipped over the top) and that they must generally be prepared by hand. In addition, they are very difficult for large groups to see.

4.7.2 Cards on easel

A medium closely related to the flip chart is the easel on which large cards are placed. These have the additional advantage that visual frames can be changed with less rustling and tearing of paper than is necessary for standard flip charts. They are, however, more expensive to make and harder to carry than the flip chart, which can be rolled. For semiformal briefings intended for presentation to two or three people, page-size, stiff cards are often a good choice because they are easy to carry in a briefcase and can be professionally displayed on a small desktop easel as shown in Figure 4.18.

4.7.3 Selection criteria

Flip charts are the best visual aid medium choice when one or more of the following conditions apply:

- A projector and screen will not be available.
- The material presented must be very current.
- The briefing is expected to lead to a group discussion.

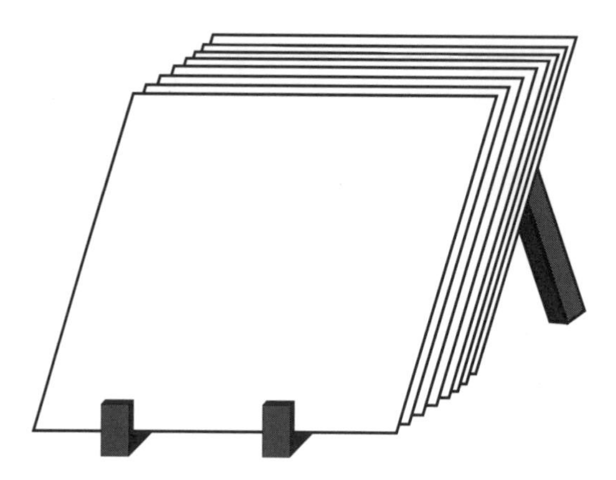

Figure 4.18 Flip cards on easel.

- The lights in the room cannot be dimmed.
- The audience is small (less than 15).
- The material consists mainly of words and line drawings.

4.7.4 Rate of presentation

Changing flip chart pages is very distracting to the audience. The speaker typically must turn away from the audience for several seconds and noisily maneuver an ungainly page over the top of an easel without tearing it. It is appropriate, therefore, to use fewer flip chart pages than you would use overhead projector slides. An average rate of approximately two minutes per flip chart page should be considered as a starting point.

4.7.5 Techniques and considerations

Techniques considered here include sequential disclosure and blank pages.

Sequential disclosure

Since the changing of flip chart pages is time-consuming and distracting to the audience, it is highly desirable to use sequential disclosure techniques to allow more time to be spent on each chart. One technique is to bring the bottom of the page up to the top and hold it in place with tape as shown in Figure 4.19. Then, as he or she mentions each subject on the slide, the briefer moves the tape down to expose it.

A second technique is to cover each subject on the flip chart with a piece of paper taped in place as shown in Figure 4.20. As the briefer mentions each subject, he or she removes and discards its covering paper.

Blank pages

Briefers should draw flip charts with dark, heavy lines so that the audience can see them easily. This means, however, that lines on subsequent pages can often be seen through the top page, which is distracting to the audience. To prevent this "see-through" problem, skip a page between each of your visuals to provide extra paper thickness.

4.8 Blackboard

The original visual aid for technical briefings and the standby for classroom use is the blackboard. Today, however, most "blackboards" are actually "whiteboards" on which pens with eraseable ink in many colors

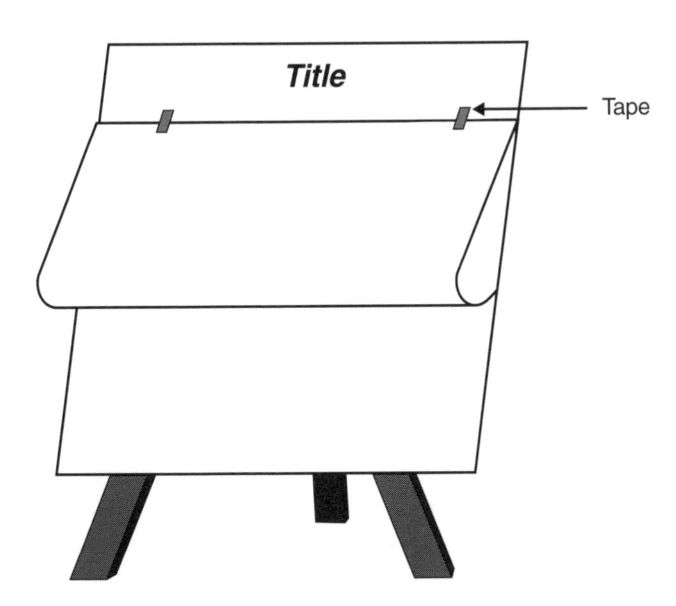

Figure 4.19 Flip chart page folded to hide material.

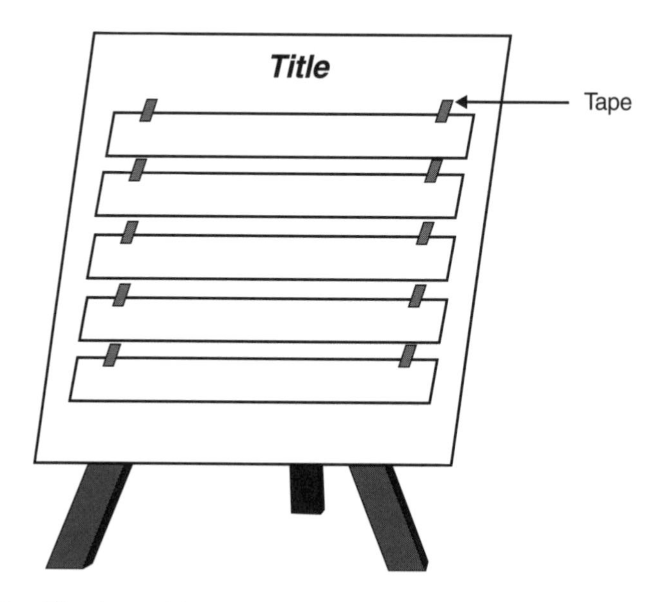

Figure 4.20 Flip chart with masks for material not yet covered.

are used. These have the advantage of better contrast for a wide range of colors so that the audience can more easily see them.

4.8.1 Advantages and disadvantages

The blackboard requires absolutely no preparation before the briefing, can be seen by a room full of people, and has infinite flexibility in adding material during the whole course of the briefing. Even when another visual aid medium is used—particularly in a classroom or formal briefing room situation—it is an excellent idea to have a blackboard available to help answer unexpected questions and side issues that arise during the course of the briefing.

The big disadvantage of the blackboard is that all material must be entered onto it by hand. Although a little material can be placed on the board before the audience arrives, most material must be added with the audience present. This requires that the briefer's back is to the audience (eye contact is lost), and it is inefficient since it takes time away from the briefing itself. Another significant problem is that some briefers have illegible handwriting, and there is no way to get typed or formally drawn material onto the blackboard.

Moreover, the chalk used on blackboards gets chalk dust onto the hands and clothing of the briefer, which does nothing for the professional image presented to the audience. Whiteboards, on the other hand, generate almost no dust.

4.8.2 Selection criteria

The blackboard (or whiteboard) may be the best visual aid medium choice when one or more of the following conditions apply:

- You are speaking in a classroom situation.
- The material presented must be absolutely current.
- The lights in the room cannot be dimmed.
- The audience is of moderate size (5–50).
- The material consists mainly of words and very simple line drawings.
- The accuracy of the words or drawings presented is not critical.
- The main material is presented on another visual aid medium.

4.8.3 Techniques and considerations

Good blackboard technique is straightforward. Be sure the board is clean when you start, and write large and legibly. If possible, prepare complex material before the audience arrives.

4.9 Handout material

Handout material can frequently fill the role of visual aids. It is particularly useful with very small audiences, informal briefings, or briefings in an office or "living room" situation where it is not convenient to use formal visual aids. Handouts are used just like any other visual aid, except that the briefer directs the audience's attention to the desired pages rather than physically changing visual frames. Written material is also useful if handed out at the end of a briefing to supplement the briefing with backup details that are too detailed or voluminous for verbal presentation.

4.9.1 Advantages and disadvantages

Handout material is easy to prepare for small groups, and, since it is usually meant to be carried away, it will make a lasting impression on the audience if it is reviewed later. Handout material can contain a great deal of detailed information, which would just be forgotten when only presented verbally.

There are two principal disadvantages to the use of handout material. One is the logistics of preparing and distributing it to large groups. The other is the fact that the audience can leaf through it while you are speaking, causing some distraction during your talk and robbing you of the element of surprise, because the audience knows ahead of time what you will be presenting later.

4.9.2 Selection criteria

Handout material may be the best visual aid medium choice when one or more of the following conditions apply:

- The briefing is informal.
- No projector, screen, or blackboard will be available.
- The material presented must be very current.
- There is a great deal of information to be presented.
- The lights in the room cannot be dimmed.
- The audience is of any size (but the logistics get tough when the group is very large).
- The material is easily reproduced.

4.9.3 Techniques and considerations

Number the pages

It is always a good idea to number the pages of a handout sequentially. This is particularly important if you are to use the handout as a visual aid, since you need to guide the audience through the pages as you cover the material. When the pages are not numbered, or worse, when there is more than one sequence of numbers (often the case when multiple documents are combined in the handout), the audience gets lost. The problem grows even worse during question-and-answer periods, when you are trying to figure out which page someone is talking about.

Quality of handout copies

Many times, briefers will be asked to supply hard copies of their visual aids to hand out to the audience. Since the hard copy will be carried away, its physical appearance will make a lasting impression on the audience. You want the quality of the copies to be good, so you should have them made from the original art used to generate your visuals. There is a temptation to just copy the slides when overhead projector slides are used, but the quality of the resulting copies is often low.

Supplementary material

Another consideration is that the visual aids will be presented along with verbal comments. However, copies will need to stand alone when they are reviewed by the audience members, perhaps months after the briefing. It is a good idea to review the handout to see if an additional page or two of explanatory material is required to make the handout useful as a stand-alone document.

One of the advantages of using a computer program to generate briefing slides is that a handout can be conveniently created as the slides are being generated. Several handout formats are typically available: two slides per page, three slides per page with extra notes beside each slide, one reduced slide per page with additional notes on the bottom half of the page, or several reduced slides per page.

4.10 Physical items passed around

If the subject of the briefing is a small object or the small object is an important consideration, the object itself can be brought to the briefing and passed around as a visual aid.

4.10.1 Advantages and disadvantages

People believe their sense of touch even more than they believe their eyes. Holding something in your hand makes a lasting impression, particularly if you know ahead of time the impression you are supposed to receive. Even if the device the audience is holding does not work, people will believe in its reality much more by being allowed to touch it than they will from any verbal description or photograph.

If the audience is not too large, the object will not hurt the audience, and the audience cannot destroy the usefulness of the object with normal handling, then it is an excellent idea to pass around an object to make a point. It must be stressed, however, that the audience must know what to look for *before* the object is passed.

Similarly, it is important to note that the physical passing of an object from hand to hand is quite disruptive if the briefer is trying to talk at the same time. Thus, the passed object should be used sparingly (only with very small groups) to make good use of the briefing time.

4.10.2 Selection criteria

Pass-around items may be the best visual aid medium choice when one or more of the following conditions exist:

- They are available in expendable or nonbreakable form.
- The three-dimensional appearance or texture of the object is important to the talk.
- The lights in the room can be left bright.
- The audience is small (less than 20).
- The object will be used as part of a talk in which other visual aid media are used to convey theoretical information.

4.10.3 Techniques and considerations

The most significant problem with pass-around items is that people have an amazing ability to break things accidentally. If the item is valuable, it is better to send around a less valuable copy of it if possible. For example, if the pass-around item is a manufactured part, try to get a similar part that is a factory reject. If only the real thing will do, then by all means pack it properly in a plastic box and *glue* the box shut.

4.11 Three-dimensional models

Seven million servicemen learned to disassemble and assemble the M-1 rifle during World War II. Almost all of them were taught with oversized models developed by the Navy for use by all services (see Figure 4.21). As the instructor removed each piece from the model and held it up, the recruits screamed out its name. The recruits could easily see the interaction of the parts and the motions required to remove and replace each. Then, they practiced. The result was that within a day or two, all of the recruits could disassemble and reassemble an M-1 rifle blindfolded.

4.11.1 Advantages and disadvantages

Even when it does not function, a three-dimensional model creates a higher level of audience interest than two-dimensional visual aids because it gives an extra sense of realism. When complex mechanical relationships or interactions must be explained in the briefing, three-dimensional models are much more efficient than drawings. To get three-dimensional information from drawings, the audience must apply extra skill and effort to relate three-dimensional drawings or sequences of drawings to the real objects or interactions they represent. For example, it is almost

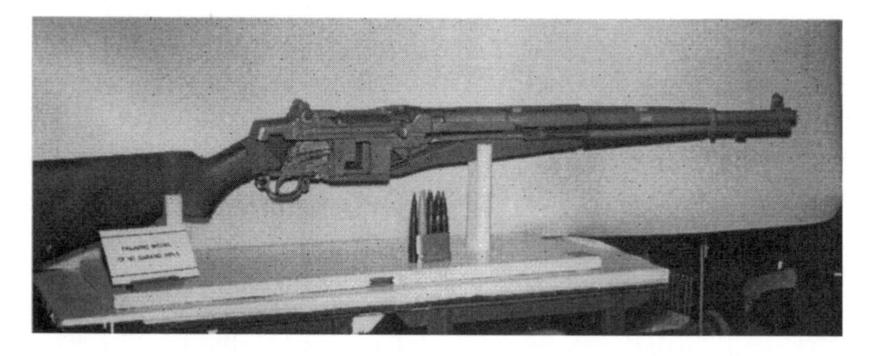

Figure 4.21 Instruction model for M-1 rifle. (Photo reprinted courtesy of U.S. Army Rock Island Arsenal Museum.)

impossible to explain clearly the operation of a gyroscope under acceleration or turning conditions without using a model.

The main drawback of three-dimensional models is that they require an inordinate amount of preparation time to manufacture. For commonly taught subjects, three-dimensional, functioning models can be purchased from school supply houses. Briefers can buy other useful but non-functioning models (for example, aircraft models) from hobby stores.

4.11.2 Selection criteria

Three-dimensional models may be the best visual aid medium choice when one or more of the following conditions are present:

- They are available.
- The subject matter is useful for a long period of time.
- Physical interrelationships are important to the subject.
- The lights in the room can be left bright.
- The audience is small to moderate in size (up to 30).
- Detailed theoretical information can be presented in another visual aid medium.

4.11.3 Techniques and considerations: getting it there

Your model will do you no good unless you can have it present at your briefing. While you are designing the model, consider how you will get it to the briefing site. This is particularly important if you must travel to the briefing by airplane. If the model is large, it should be easily disassembled with small tools so it can be carried, checked, or shipped. Carry-on items must fit under an airplane seat. Check with the airline for maximum dimensions; each has a different specification. Maximum carry-on dimensions of 8 inch × 13 inch × 21 inch are the norm, but not always acceptable. There are also limitations on the size of packages checked as luggage, although they are not as severe as those for carry-on items. It is good practice to design the model to fit into a standard piece of hard-sided luggage. The luggage provides some protection, has a handle (and maybe wheels) to help move it, and you know the airline will not give you any grief about checking it as luggage. If you plan to check a model as luggage or ship it ahead, it must be well protected enough to withstand the handling it will receive. You feel very lonesome when you discover two hours before your briefing that your model arrived badly broken.

4.12 Demonstrations

Demonstrations do not always involve full-scale operation of actual hardware or performance of actual processes. Use of two- or three-dimensional models, software simulations, time-compressed operations, and idealized representations of phenomena or processes are often the most effective demonstrations.

Figure 4.22 shows an example of a very powerful demonstration. The purpose of the briefing is to convince the operators of rental equipment companies to add ground-fault interrupt (GFI) devices to the equipment they rent to the public. In the demonstration, a large humanoid model

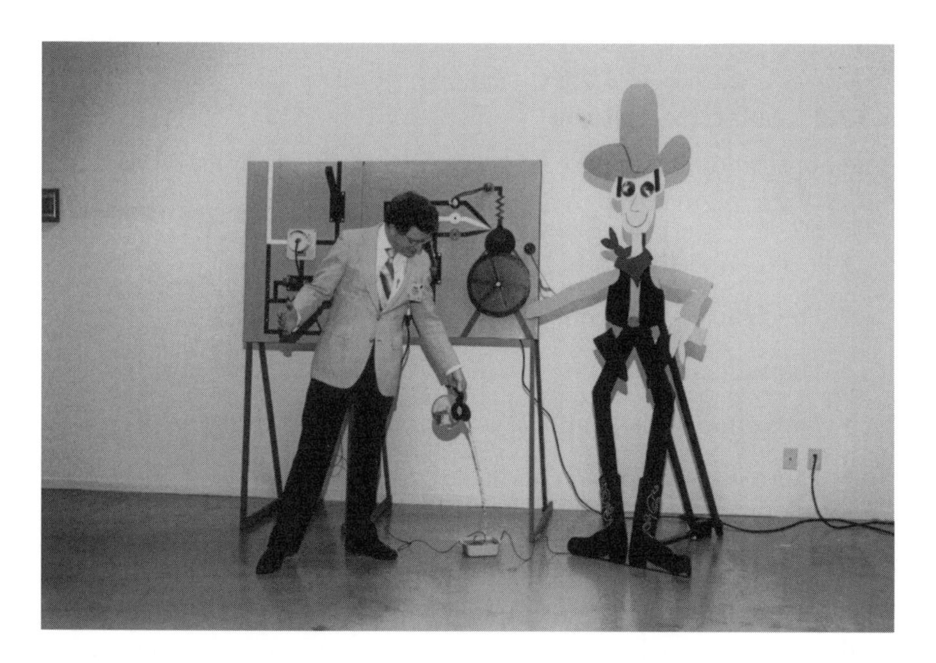

Figure 4.22 Electrical safety demonstration. (Photo reprinted courtesy of Sotcher Measurements, Inc.)

named Alfred is hooked up to an electrical current, with and without the GFI device in the circuit. Without the GFI, Alfred reacts in a most graphic manner to demonstrate that he is being electrocuted. Since it is a cartoon character that is being electrocuted, the event is funny rather than tragic, so there is humor to add to the impact of the lesson being taught.

4.12.1 Advantages and disadvantages

"Seeing is believing," so a demonstration is an extremely effective way to make a point. Demonstrations are often used as part of a series of briefings and can be included within individual briefings if time permits and the goals of the briefing are met.

The biggest problem with demonstrations is Murphy's Law: "If anything can go wrong, it will." Demonstrations that do not work are embarrassing to the briefer, weaken arguments, and waste immense amounts of time. An effective demonstration requires a great deal of preparation time, both to assemble the necessary equipment and to be sure that the demonstration will proceed as planned and come to the desired conclusion.

4.12.2 Selection criteria

A demonstration may be the best visual aid medium choice when one or more of the following apply:

- It is logistically practical.
- The lights in the room can be left bright.
- The audience is of small to moderate size (up to 30).
- The action of the demonstration is important to the subject.
- There is plenty of time to make sure it will work.

4.12.3 Techniques and considerations

It is a universally held truth that things will go wrong in a demonstration. Further, any informal survey of experienced demonstrators will uncover the conviction that there is a direct correlation between the importance of a demonstration's succeeding and the likelihood that it will fail.

Nevertheless, demonstrations must be made, and they can have a high rate of success with proper planning. The following steps will help you sidestep potential problems:

1. Consider what will be available at the demonstration site:
   - What power will be available?
   - What will the environment be like (e.g., heat, light, and radio frequency interference)?
   - What must be provided by the sponsors?
2. Make your demonstration independent of outside support as much as possible. If you can design it to run on internal batteries and to require no additional support equipment or inputs, there will be fewer things that can go wrong.
3. Design your demonstration so that it will arrive at the demonstration site in operational condition.
4. Take *everything* you will need with you:
   - Power adapters;
   - Extension cords;
   - Replacement parts, especially unique items;
   - Repair tools;
   - Manuals.
5. Consider ahead of time what might go wrong and what you will do to correct such problems.
6. Get there early and try the demonstration at the actual demonstration site without the audience present and with enough time to fix the minor things that *will* go wrong.

5

Creating Visual Aids

THE HARD PART OF CREATING visual aids is determining what goes onto them, but what the audience sees may not reflect this selection process and decision making.

This chapter describes ways to generate quality visual aids in each of the media discussed in Chapter 4. It includes the generation of each type of visual aid with and without the help of a professional graphic arts department.

5.1 General requirements

Whether you prepare your own final visual aids or have them produced by a graphic arts group, a methodical approach will allow you to do a better and more efficient job. Naturally, when you are doing the entire job yourself, you can be more informal than if you were to pass the information to someone else, but it is still a good idea to make careful notes so that you will not confuse yourself later in the process.

This process is presented as though it were a logical progression of events, with each step completed before the next is started. In the real world, it will probably be a reiterative process in which changes must be made at fairly late points. Still, a methodical approach will allow you to minimize the changes required and help you find mistakes early in the process. As you get further and further invested, changes are harder to make, so it is wise to find the errors early.

This is a case in which new presentation software can be very helpful. You can start by generating an indented outline and then convert it to a presentation format. The result is a set of bullet slides. Your main headings will be slide titles, and the subordinate items will include bullets, first-level subordinated bullets, and second-level subordinated bullets.

5.1.1 List of visuals

Carefully review the list of visual aid frames you made as part of the design of the briefing to be sure that it is complete and that it presents your material in the order in which you want the audience to see it. This is by far the easiest point at which to add, delete, or rearrange material.

At this point, you should also check to see if some of the material is already available to you in final form. It is particularly desirable to use expensive items (such as color photographic overhead projector slides) in existing form if possible. You can also import slides from any other presentation (that is held in memory on your computer) and place them wherever you like. Sometimes, a minor change in the organization of the visuals will allow you to save yourself a great deal of trouble by using some existing material.

5.1.2 Overview of visuals

After reviewing your list of visual frames, it is an excellent idea to make a way to scan your visual aids. Not all the substance of the material is apparent in the printed list of titles. The visual coherence of the presentation is also important. With a visual overview, you will be able to see the subtleties of visual flow and how each frame fits into the big picture. This kind of an overview will give you more information than even making a dry run through the full set of final visuals (when they are eventually ready) because then you will just see one visual at a time, while now you can see them all at once. Together they will indicate the *trends.*

If there are just a few visual aids, the rough drafts can simply be laid out in sequence on a table so you can see them all at once. Some people like to make small sketches of all of the visual aids on a single piece of paper, as shown in Figure 5.1.

For larger scale or more formal briefings, more elaborate schemes are in order. An excellent overview technique for the larger briefing is to make a sketch of the material that will be on each visual on a single 3 × 5 card. Then, the cards can be laid out in sequence on a large tabletop, arranged in a commercially available planning board, or pinned to the wall. Whatever technique you choose, it should allow you to add, delete, or rearrange material easily. Moreover, it should, of course, allow you to see all of your visual frames at once.

An overview technique is particularly important when you are planning to display multiple visual aid frames at one time (for example, 35-mm slides with multiple screens). You need to review everything that the audience will see at one time to be sure that everything required is present and to avoid redundancy.

If you are using a presentation software package, you can bring up a "slide sorter view" that shows miniatures of all of your slides. Thus, you can see several slides at once and scan easily through the whole presentation, rearranging the slides as you see fit. You can also add new slides anywhere in the sequence. These can be new bullet slides, crudely drawn draft graphical slides, or existing slides imported from another presentation.

5.1.3 Making rough drafts

A good rough draft of a set of visual aids will be complete and legible and will indicate exactly what the briefer wants to appear on each frame. When you plan to make the final slides yourself, there is a great temptation to either skip the rough draft step or to do a very cursory and perhaps inaccurate job of it. While complete rough drafts obviously are not as critical in this case as they would be if an artist were doing the finals, there are still annoying, time-consuming pitfalls that a good set of drafts will avoid. First, it is often possible to have someone else (for example, the boss) review your presentation visuals in draft form to be sure that you are

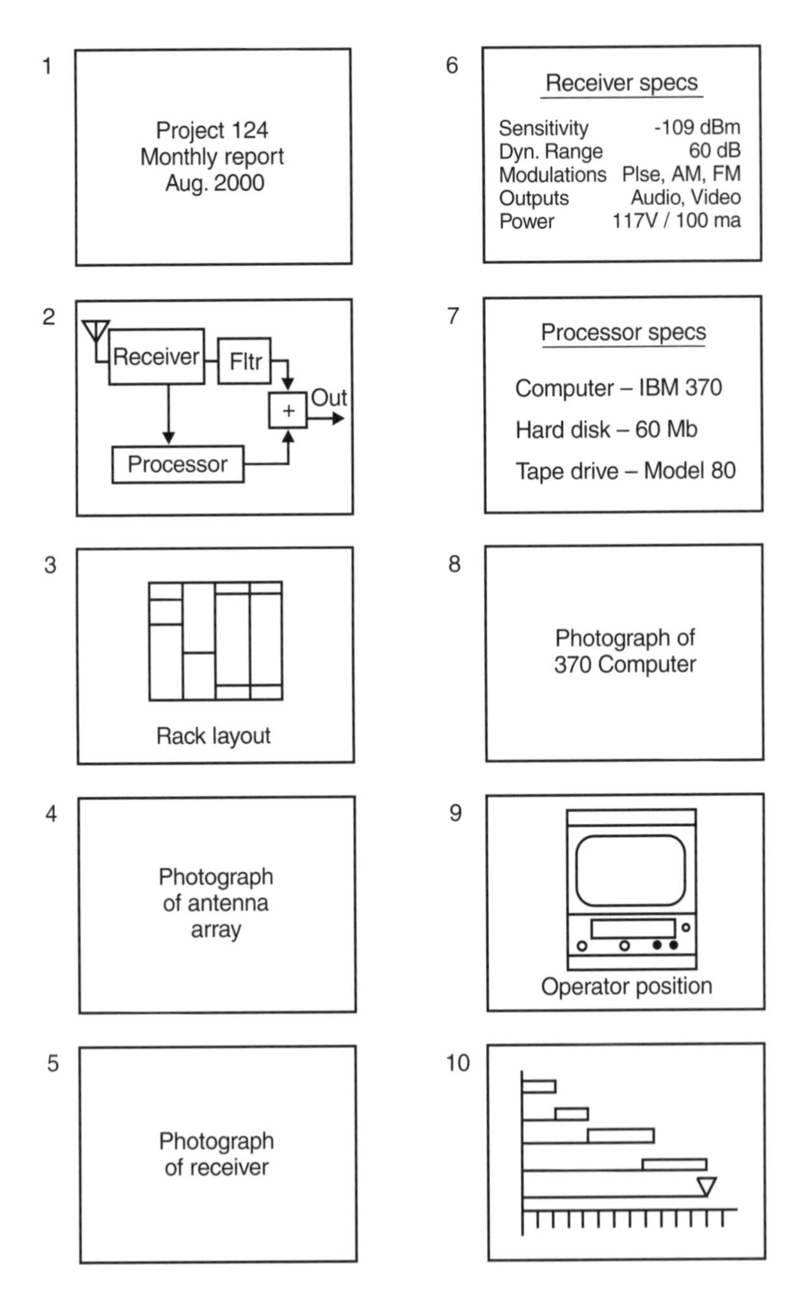

Figure 5.1 Sketches of visual aids.

briefing what the boss thought you were to brief. Second, if you are preparing a large number of visuals, you will avoid confusing yourself during the review and production process by preparing a complete and accurate set of rough-draft visuals. Artists say that they have trouble with even simple spelling when they are concentrating on the artistically oriented task of producing presentation-quality text.

The rest of Section 5.1.3 is written with the assumption that you will be handing your rough drafts off to an artist for development of the final visual aids. If you will be producing your own final visuals you may want to skip some steps, but please note the above cautions.

It is preferable to develop your visual aid drafts using a dark pencil (2B lead is nice) and a large, good eraser. If you use a pen, you will either find yourself starting over many times as your ideas develop, or you will scratch out material, making the final draft confusing. In general, paper is cheaper than time, so do not try to correct a rough draft to the point that it is hard to read. Redrawing the draft when you have figured out what you really want will save you both time and aggravation later in the process. Furthermore, in the name of every eye-strained artist in the world, please make your drafts neat and use lettering that is large enough to be read comfortably.

When making drafts for visuals that will use multiple colors, you must indicate what colors you want and how you want them used on the slides. Carefully indicate colored lines and colored areas. It is an excellent idea to use a set of colored pencils to show directly how you want the colors used. If you do not have them, or cannot use them for some reason, be sure to indicate your intentions with side notes. These rules also apply to special lines (for example, a dashed line or wide line) and to shaded areas in drawings.

Draw your figures as accurately as possible and show dimensions where appropriate. Even if the overall size is not important, the dimensioning will ensure that objects in the final artwork are the proper shape. If your draft includes an exact scale drawing, note that fact and include the scale in a side note.

Be sure to number your visual frames in some consistent way so you can keep track of them during the production process. In addition, carefully label overlays. Indicate the slide to which the overlay will be mounted, the nature of the mounting, and the purpose of the overlay in the scheme of your briefing. When you want to use multiple overlays on a single slide, it is a good practice to make a side note with a sketch to show how they are to be mounted.

If you have a software version of your briefing (often called a "soft copy"), it is an excellent idea to give the artist both a disk and a hard copy printout. The hard copy keeps the artist from getting lost in the electronic version and gives you the opportunity to add written clarifications of your intentions. If your presentation contains some slides that are only available in paper form, you can stick copies into your hard copy and note how they are to be incorporated. Working from your soft copy will often allow the artist to rearrange as needed and modify most or all of your material into the final form. Even if the artist decides to redraw your graphics, the electronic version may simplify the task.

5.1.4 One final review

Before you start generating final visual aids, it is a good idea to review your list of visual frames and the draft copies of the frames. Take a moment to flip through your visual frames in sequence to be sure that you have all of the frames you intended to generate (without duplicates) and that they are compatible. Ask yourself whether they are in the proper order, delivering information to the audience in the optimum sequence. If the audience needs to know fact A before it can appreciate fact B, then be sure the visual on fact A comes before the visual on fact B. Make any changes in order and add or delete visual frames as required.

When you have made a full draft of your visual aids, it is time to prepare the actual visual aids you will use in the presentation. The rest of this chapter covers ways to produce final visuals, with or without the assistance of a graphic arts support group.

5.2 Using the services of a graphic arts department

In general, most of us are willing to take any help we can get. An organization's graphic arts group is skilled and equipped to turn out professional visual aids. If you have authorization (and funding if necessary) to use the group and it can produce your visual aids to meet the schedule you require, by all means use it. If your organization does not have such a group, there are independent visual art firms that can provide the same service on a fixed price or hourly cost basis. Everything stated in Section 5.2.1–5.2.5 about working with an in-house graphic arts group applies equally to an outside supplier of these services.

5.2.1 Working with an artist

When a graphic arts department is creating visual aids for you, you can expect to be working with an artist who knows absolutely nothing about the subject of your briefing. This is, of course, not always the case. Sometimes there will be one or two people in a large graphic arts department who at least know the terminology, but those people will be in such demand that you may never get to them. Expect, therefore, to work with a "typical" artist with the following characteristics.

- Does not understand the technical subject;
- Does not know the terminology;
- Is expert in the mechanics of generating visual aids;
- Understands the subtleties of color combinations;
- Understands the subtleties of composition and layout;
- Knows all of the standard operating procedures for visual aid composition and layout in your organization;
- Has the tools and materials to make a first-rate product;
- Is expert in the handling of artists' tools;
- Is very busy on several jobs at once.

What you see above is the profile of an expert in the artistic field who is not expert in your field. By taking the time and effort to establish a good relationship, you will get a better set of visual aids in less time and at less cost. Like a relationship with any expert in a field other than your own, a good relationship with an artist involves mutual respect and the sharing of expertise in terms that the other person can understand. Here are some general guidelines for working effectively with an artist:

- Start working with the artist early, to learn what media are available within your organization and what the artist needs from you to produce the end product you want.
- Take the time to tell the artist a little about the technical subject matter you will be covering in the briefing. Also explain what you are trying to accomplish with the briefing and how you plan to handle the presentation of the material. An artist who understands the situation will help you make your briefing successful.
- Tell the artist how you plan to handle the presentation.
- Be careful to input your drafts in as close to the final form as you can manage, with all the words legibly spelled out and all of the drawings laid out as you want them to show on the page.
- Be sure to mark any important dimensions on drawings you turn in to the artist, unless the draft is drawn accurately to scale and the artist knows it.
- When there is something subtle about the material on a visual frame, make a peripheral note to explain it completely.
- Keep an exact copy of what you have submitted to the artist so you can discuss minor corrections and clarifications over the phone.
- Give the artist your original sketches and make copies for yourself. Any copying process can make something less accurate or harder to read, and you want the artist to be working from the most accurate material possible.
- Number your visual frames and turn them in to the artist in the correct order.
- Make your draft neat enough to be read *easily* (typed or printed letters, dark pencil lines on drawings) by the artist.
- Listen to what the artist has to say about layout and media.

- Give the artist as much flexibility as you can in organizing the material without making the material inaccurate. The artist knows more than you do about how to make things look good. Given a free hand, the artist will make minor changes in layout that will make your visuals easier to read. However, as experienced artists in technical situations know, a seemingly innocuous change made for artistic reasons can sometimes make a visual aid technically inaccurate.
- Proofread every stage of the material generated by the artist early in the process. Remember, the artist does not know the terminology and, therefore, might not be able to detect an obvious typographical error (whether made by the artist or by you) in its context.
- Ask questions about color combinations and layout before you input your final draft material. This will allow you to generate your drafts with an understanding of what the final product will look like.
- Be honest about schedules. Tell the artist when you really need your material and pass on any schedule-change information. If you get the reputation of always demanding artwork long before you need it just to be safe, the art department will soon have you "categorized" and will not take your schedule seriously. Someday, that will come back to haunt you when you really do not have any extra time.
- Get your material to the artist as early as possible. A little extra time may give the artist the flexibility to use different tools or processes to make the final visuals look better. It will also allow the artist time to go back and take another look at the visuals to be sure that there are no typographical errors. If either you or the artist has made a mistake, extra time in the production process will make it possible to make changes before your presentation.

Just like any professional, an artist wants to turn out an excellent product and wants its final implementation (in this case, your briefing) to be as successful as possible. By following as many of the above guidelines as you can, you will give the artist the strongest chance to do his or her best job for you.

5.2.2 Understanding the medium

To create a satisfactory set of visual aids, it is necessary to think out how they will be used. In the design of your briefing, you determined not only what you were going to cover, but generally how you would present it to the audience. The limitations of the media were considered in that process.

5.2.3 Specifying layout

Specifying the layout for your visual aids is a fairly simple process but should be done before any final art is generated. Your organization may require compliance with visual aid layout standards. If so, be sure that the artist is familiar with the general requirements as applicable to the type of briefing you plan. Examples of required layout constraints include the following:

- A need for the organization's logo on every visual frame, in a specified place;
- A mandatory colored band at the top of visuals with titles of visual frames;
- An assigned control number for each visual frame.

In addition to any required layout items, you may want to add something to the standard layout of all the visuals in your briefing to provide extra unity. Examples of such layout items are listed as follows:

- Use of a specific type font or type size;
- Horizontal versus vertical page layout;
- Less than full frame visual aids.

5.2.4 Finding photographs

There are many sources for photographs for visual aids. Accordingly, you should make a reasonable search before deciding to take a complete set of new photographs. The first place to check is the organization's files. Most graphic art departments have an extensive set of files that are very useful if you understand how the filing system works. The manager of the department is the best person to ask for an explanation of the system. Usually, a file number will be associated with every photograph taken by the department's photographer or otherwise obtained for the files. In a well-run group, each negative will be stored in an envelope marked with this number, and all prints or slides made from that negative will be marked with the file number.

If you know the file numbers for the pictures you need, your problems are over. Just request prints or completed visuals from the desired file number negatives. Most art departments have *very firm* rules against releasing the negatives to anyone, so do not even try to get the negatives unless you have an awfully good reason and are ready for a long, hard battle.

The problem with file numbers is that you almost never know them and probably cannot get the pictures from the files without them. You just know what you want the picture to show. There are several ways to get the numbers:

- If you see a print or a slide made by your graphic arts department, look for a file number on it.
- If you see an appropriate photograph in a report, note the report number and go back to the group that produced that report to find the original, which should include the photo file numbers.
- Determine the project (or program) associated with the subject area in which you need photographs. Then go to the graphic arts department and ask to see file photographs for that project. The file number system is often based on projects, or at least has a cross-reference to projects.
- If all else fails, go to someone in the art department who has been around for a number of years and ask how to find photographs in the general subject area. This should be the last alternative because it can be very time-consuming unless you have some leads to follow—if your organization has been around for a while, there are probably thousands of photographs in the files.

In any case, when you have the file number of a photograph that is close to what you want, ask to see file copies of the rest of the pictures in the same general subject area. Commercial photographers normally take several photographs from different viewpoints or under slightly different conditions while they have their equipment set up, and you will want to get the photograph that is best for your needs.

Photographs are often available in digital form, having been taken with a digital camera or scanned from a paper photograph. Most art departments can accept any common digital figure format and convert it to their required format for incorporation into the final product. However, be sure to ask what format they prefer (for example, JPEG) so that you can choose the most convenient form whenever you have a choice (which you usually do).

5.2.5 Scheduling and budgeting work

Be sure to discuss the schedules and budgets for your work with the art department very early in the process. This may limit the media you can use and the nature of the artwork you can afford to have generated. It is far better to learn ahead of time that you cannot afford first-class, professional artwork than to get halfway through the process and run out of money or time.

Once a budget and schedule have been agreed upon, manage the work the art department is doing for you just as you would manage any other technical work. Get agreement on intermediate progress goals and be sure that they are met. Be sure that the product is being developed properly. Cost, schedule, and technical content problems are much easier to correct if caught early.

5.3 Generating visual aids by yourself

If a graphic arts department is not available to you, or if you do not have the time or budget to use the department, you can still have a first-class set of visual aids by generating them yourself using commonly available materials and equipment. You do not need to be an artistic type to make professional-looking visual aids. If you work carefully and neatly, your visual aids can look professional.

The balance of this chapter describes ways to generate your own visual aids in the following media:

- Overhead projector slides;
- 35-mm slides;
- Moving media;
- Flip charts and display cards;
- Three-dimensional models;
- Direct computer projection.

5.4 Overhead projector slides

With the invention of the office copier and transparency film that can be used with it, overhead projector slides became the most convenient visual aids for a briefer to prepare without the benefit of an artist.

5.4.1 Using a copy machine or computer projector to make slides

Before the early 1960s, the only ways to make prepared slides for overhead projectors were the expensive and lengthy photographic process or an Ozalid™ process that required special equipment and strong-smelling chemicals. Self-prepared slides were typically drawn with grease pencils. Special pens to write on acetate were not widely available.

In the early 1960s, Minnesota Mining and Manufacturing Company (3M) marketed a series of heat-sensitive films to allow the direct production of overhead projector slides on a thermofax machine, which was widely available in offices at the time. For the first time, individuals could make their own overhead projector slides right in the office from black-on-white material.

The thermofax material was available in four colors and in positive or negative image. The negative image process was particularly attractive because it created bright, colored letters against a dark background, making text very easy to read even in a brightly lighted room.

In the early 1970s, the thermofax was replaced by the Xerox™ and other electrostatic copiers as the most common office copier, and several companies started making film that would go through the machines like paper and make black-on-white images for overhead projector slides.

Laser printers can print black and white overhead projector slides from any type of computer file. They use plastic stock similar to that used by copy machines. However, you need to be sure that the stock is compatible with your printer to avoid jams or worse. Since the laser slide material can be more expensive than the equivalent copy machine material, you may be able to save money by printing a paper copy and then making slides on a copy machine.

Ink-jet printers can print colored slides directly from computer files. Again, be sure that you use slide stock that is designed for your printer.

5.4.2 The process

Film for overhead projector slides is packaged in boxes of 100, but it is sometimes available in single sheets or small packages from stationery stores. You must choose the proper film to match your office copier. The copiers for which each film type is appropriate are listed on the box. Two weights of material are available from most manufacturers. The lighter material is suitable only for use on slides that will be mounted on slide frames. Heavier material is specifically for "frameless slides" that are not mounted on slide frames. Electrostatic copiers are limited to black images (on clear or colored background) because of the nature of their copying process. However, color copies are also available. Be sure to use the right film to avoid smearing if you choose an ink jet printer.

5.4.3 Techniques

The following general techniques apply to the generation of overhead projection slides with most copiers. For good-looking slides, use an original with good contrast. Anything that will copy on the machine will go onto the slide, but higher contrast is important since the finished slides are often not completely transparent. Sections 5.4.5–5.4.7 discuss specific original generation techniques.

- With most copiers, you can place several sheets of film on top of the paper in the paper tray and make several slides until the slide material runs out. However, the plastic slide material has a tendency to stick together and feed in multiple from the paper tray. For both of these reasons, you may find it worth the trouble to feed the slide film sheets one at a time.

- Some copy machines seem to heat up when used with overhead slide film, so it is good practice to pause for the machine to cool down after a few slides.
- The slides come out of the copier hot, and if allowed to curl will take a permanent set that will make them difficult to use, particularly if they are not mounted on slide frames. It is a good idea, therefore, to catch each slide as it comes out of the machine and lay it on a flat surface to cool.
- Many copiers make copies on film, which looks different from those made on paper, so experiment with the density adjustment when making slides. You want good contrast, but if it is high enough to show background shading, the viewability of the slide may be significantly reduced.
- It is important to ensure that the glass on the copy machine that holds the original is clean; otherwise any specks or smears will be reproduced on the slides.

5.4.4 Enhancing copy machine slides

There are two convenient ways to make copy machine–generated slides look more professional—by adding selective color and shading. Very light gauge, translucent, self-sticking acetate material is available in many colors. It can be cut to shape with a sharp knife and used to add color to any area of the slide. One common use is to provide a colored title block area at the top of the slide. If the same color is used in the same title area of each slide in the briefing, it will add unity to your set of visuals.

Artists use self-sticking, transparent material with various dot patterns to provide shading. This is called screen and is cut to shape and placed onto the finished material in the manner described for colored material. (Please note that screen can also be applied to the original art and will copy very nicely onto the slide itself.) A very effective technique is to place the colored or shaded areas on overlays so that parts of a drawing can be highlighted as the talk proceeds.

In addition, several companies make pens specifically designed to write on plastic materials. They come in a wide range of colors, in at least four widths, and in permanent or water-soluble ink. They are available in most stationery stores either as individual pens or in sets of assorted colors. These pens are ideal for adding line type material to slides. As pointed out in Chapter 4, they can also add material to slides for dramatic effect during the presentation itself.

Although the pens can be used to color in areas of slides, they are less satisfactory than the acetate sheets because the ink dries very quickly as it is applied. It is smeared by subsequent pen strokes, making the individual pen strokes quite obvious and creating an unprofessional appearance.

5.4.5 Typed, drawn, and lettered slides

Although anything that will reproduce on the copy machine can be used to make a slide, the contrast must be good to make a nice-looking slide. The original can be drawn with a dark pencil or, better yet, with a black felt-tip pen. Written material can be typed, lettered with a template-based system (such as Leroy), free-lettered with a black pen, or produced using dry transfer (usually called rub-on) letters.

Naturally, most originals these days are created on a computer. Text can be set to any desired size, and graphics can be generated with graphics software. (This is discussed in detail in Chapter 6.) Still, it is good to consider the other techniques because they sometimes provide the best solution.

Remember that the size of the type must be appropriate to the overall size of the viewing area and that there should not be too much material on any one slide (see Section 4.1). When typing originals for slides, it is a good idea to use the largest type font available. Orator type, which is used to type speaking notes, is normally the largest available and makes very readable slides. Eighteen- to 24-point type size for computer-generated slides is normally best. Plain type fonts generally make more satisfactory slides than those from fancy type fonts.

When making drawings for slide originals, remember that small "wiggles" and variations in line width will be much more obvious on a large screen than they are on a normal size page. You should use a straight edge for all lines and templates for repetitive figures whenever possible.

Appropriate templates are available for drawing computer program flowcharts, electrical schematic diagrams, and many other types of technical drawings. Typing the word material onto the slide after the line art is in place is an excellent technique for generating readable and attractive slides.

Drawings made with PowerPoint or other drawing software will always yield very viewable slides as long as line width and type fonts are properly selected.

5.4.6 Slides from photographs

A copier will make overhead projector slides from photographs, but most copiers will not do a good job unless the prints are screened. This process is called half-toning. It creates a dot pattern on the photograph to allow the copier to reproduce gray tones. The problem is that an electrostatic copier, like most printing processes, sees everything as either black or white and will force gray tones to one or the other, destroying much of the information in the photograph. A screened print has no true gray areas, only patterns of completely black dots of varying size and spacing that appear gray to the human eye.

Some copy machines also have difficulty reproducing large black areas. The center of the area fades out in the copy. Screening eliminates this problem as well.

Photographs taken from newspapers, magazines, and most printed reports are already screened, so they can be used directly to make overhead projector slides on a copy machine. The quality of the slides made in this way will not ordinarily be as good as slides made from large screened prints and may be considerably worse than slides made by direct photographic processes. Photo-processing software can make a halftone from a scanned or digitally input photograph.

5.4.7 Computer-generated graphics

A hard copy printout of computer-generated graphics and/or text is an excellent original for making slides in a copier. To generate the necessary contrast, use the best resolution available from the printer. If you use a laser printer, be sure that the cartridge is not at the end of its life (i.e., making light streaks through the page). If you are using a dot matrix printer, be sure that the ribbon is new.

Slides are usually produced from a presentation program such as Microsoft's PowerPoint, but can also be made from word-processing, spreadsheet, or other programs. Most programs allow wide flexibility in the selection of type fonts and sizes and in all of the elements of graphics. Section 5.9 discusses some tricks to simplify the generation of professional-looking slides with any type of graphics software package.

5.4.8 Pen on acetate

Once in a while, you really need a set of slides and have no copy machine available. Consider the following real-life example. You are in a hotel room on the other side of the world and have four hours to prepare a set of slides for an important briefing that you just learned you must give.

When nothing else is available, you can make a very acceptable set of slides by drawing on blank sheets of acetate with special pens (described in Section 5.4.4) that are designed to write on plastic.

5.4.9 Photographic-process overhead projector slides

In addition to all of the techniques described in Section 5.4.1–5.4.8, you can also have overhead projector slides made by photographic processes. Black-on-white originals can be converted into "film positives" by any photo lab. You can have full-page-size transparencies made from any colored or black-and-white negative. The photographic processes are expensive, but they yield very high-quality slides.

5.5 35-mm slides

When your briefing involves photographs of equipment or people performing processes, 35-mm slides are the visual medium you can most easily prepare on your own. Simply use a 35-mm camera with the required extra equipment (such as a flash gun, tripod, or cable release) and take the pictures you require.

5.5.1 Pictures of equipment

There are many good books about the selection of camera equipment and its proper use for taking pictures of people and scenes, indoors or outdoors, so picture taking will not be covered here. However, there are a few special considerations when taking pictures of equipment. First, the pictures of the equipment should be made interesting. Ways to do this include the following:

- Taking pictures from angles other than straight on;
- Trying to show the equipment in operation (with lights flashing, etc.);
- Showing people operating or maintaining equipment;
- Showing equipment in its operational setting rather than sitting on a workbench;
- Taking a series of pictures to give the audience a feeling of "being there."

When taking flash pictures, you need to be careful of reflections of your flash from shiny surfaces. Holding the flash a distance above the camera or shooting pictures at an angle that is not straight onto any shiny surface will avoid this. Figures 5.2–5.4 are a series of photographs that show increasing detail of a piece of equipment in a laboratory setting.

5.5.2 Slides from written and drawn material

To create slides showing written and line graphic material, the easiest process is to take 35-mm pictures of artwork on which the appropriate material is presented.

Copying stand

For convenience when making these types of 35-mm slides, you may wish to purchase, rent, borrow, or improvise a copying stand for use with a 35-mm camera. Figure 5.5 shows a common commercial model. For one-time use, you can also rent a commercial unit from a full-service camera store. The copy stand is designed to hold the camera in proper position

Figure 5.2 First slide in series. (Courtesy of Zeta, Inc.)

to photograph your artwork. Depending on the camera used and the size of the original art you are copying, it may be necessary to use a close-up lens to focus the camera close enough so that the desired material fills the 35-mm slide viewing area. An alternative approach for occasional use involves mounting the camera on a tripod that is adjusted so that the lens is parallel to a tabletop as shown in Figure 5.6.

Lighting

You need proper lighting on the original art, particularly if you are using color slide film. The most consistent, satisfactory solution is to buy, rent,

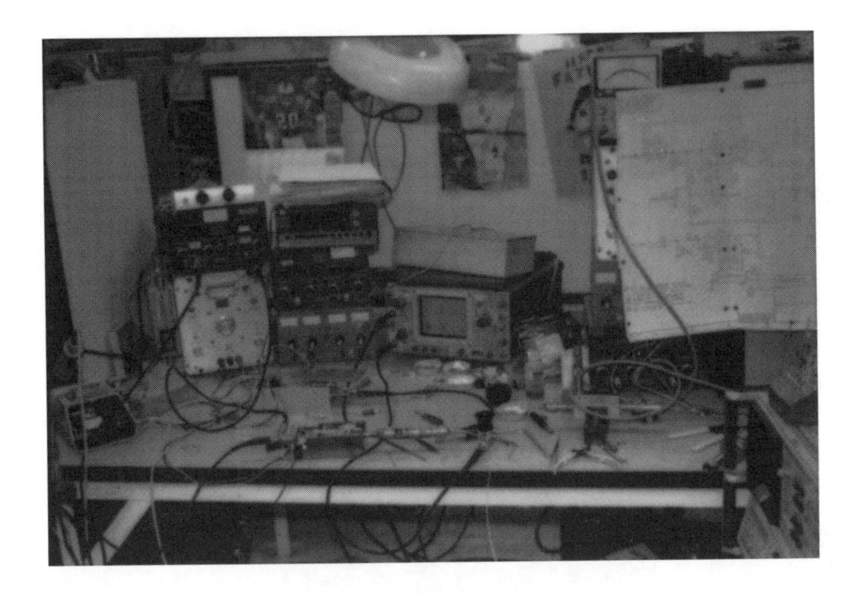

Figure 5.3 Second slide in series. (Courtesy of Zeta, Inc.)

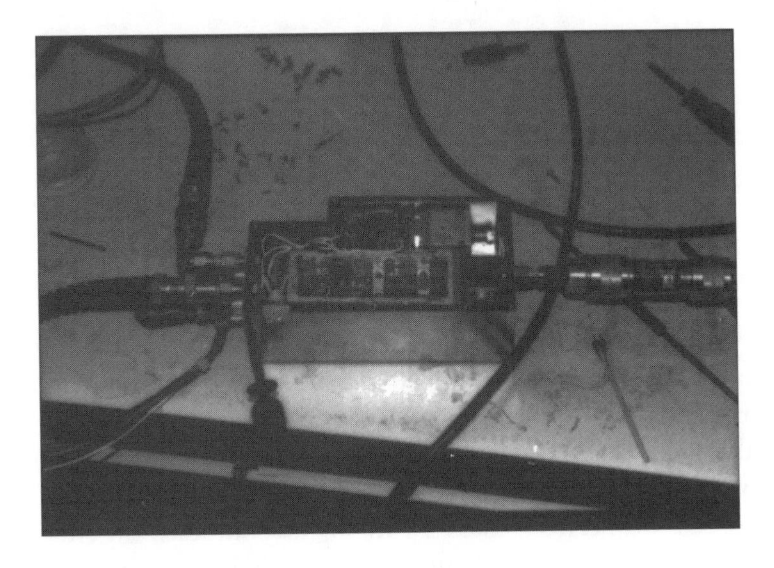

Figure 5.4 Third slide in series. (Courtesy of Zeta, Inc.)

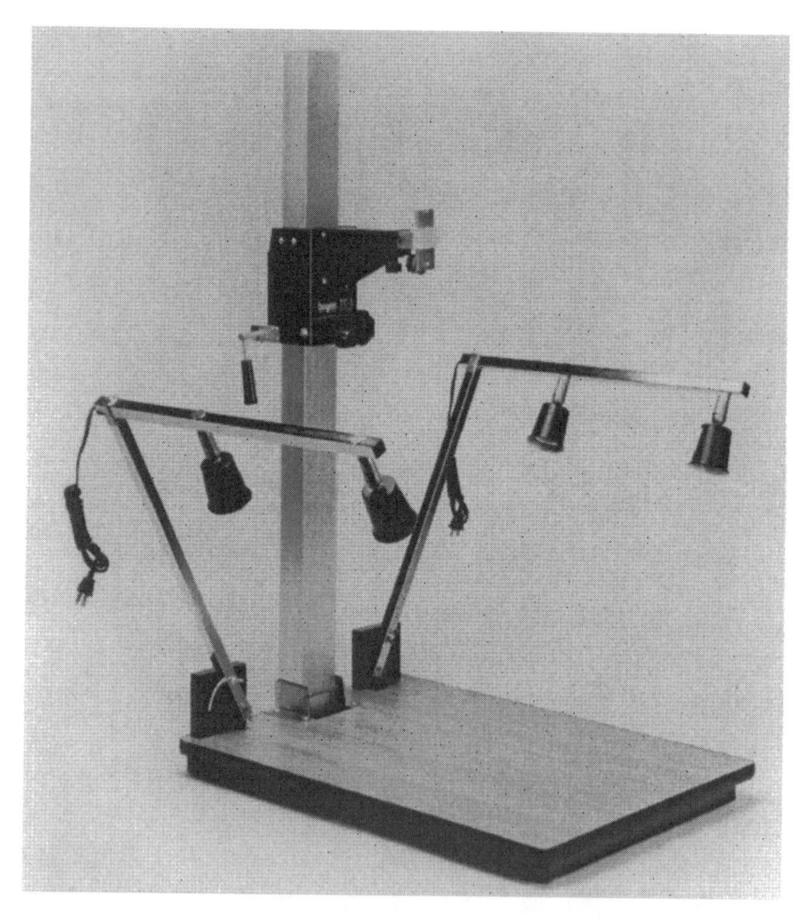

Figure 5.5 Commercial photo stand. (Reprinted courtesy of Bogen Photo Corporation.)

or borrow a set of at least two photo-flood lights. With a specified filter on the camera, true color slides can be produced by adjusting the camera's shutter speed and f-stop.

If the elements cooperate, an alternative to the use of floodlights is to move the copying operation outdoors to an area of bright shade. When the camera is properly adjusted for the light conditions, the slide colors will be true without filters. Figure 5.7 shows a 35-mm slide being shot from line art using natural light. For black-and-white film, the light color is not so critical, but there must still be sufficient light, properly located to

Figure 5.6 Tripod adjusted for making slides from artwork.

get proper contrast with the film used and to be free of distracting reflections from the original art.

Original art

There are many ways to create original art for 35-mm slides.

- You can draw text and figures with a felt pen on a large pad of paper.
- You can write and draw the desired information on a whiteboard with special pens. You can also use magnetic letters on the same board, which is usually made of steel.
- You can generate your graphics with a computer as described above for overhead projector slides.
- You can arrange letters on a commercially available title board.

Figure 5.7 Making slides from line art using natural light.

- You can create a very interesting effect by arranging plastic letters on a piece of cloth. Graphs can be created with a string and pins along with the plastic letters.

5.5.3 Modified negative slides

The sequential disclosure technique for 35-mm slides described in Section 4.5.5 can be created easily by use of identical copies of a negative of a black-on-white original drawing. To do this, shoot a number of identical black-and-white pictures of the full drawing or have a number of copies of a single negative made by a photo shop. The negatives are mounted in slide mounts just as though they were regular 35-mm slides.

Unfortunately, in order to get good bright letters and a background that looks black when projected, it is necessary to shoot with lithography film. Since this film is generally sold in page-size (or larger) sheets, it requires commercial lithography cameras. A trick to decrease the cost per slide is to lay out a large number of slides on a piece of art board. An easy way to work the layout is to draw a number of adjacent rectangles 2.75 inches high by 3 inches wide on the art board. Place the material to appear on each slide within a 1.75 × 2.75 inch area centered in each rectangle (as shown in Figure 5.8). Then, have the board shot at 51% by a commercial print shop that has a layout camera. The lithograph negative can then be cut up along the rectangle lines and mounted in standard 35-mm slide frames. If you make a layout of seven rows of slide originals with six in each row, the camera will fit 42 slides neatly onto one 8.5 × 11 inch sheet of film for maxium cost efficiency.

Use the special pens designed to write on plastic to modify the slides as required to generate a set of slides functionally equivalent to those described in Figure 4.14(a–e). Block out undesired portions of the drawing with a black pen and color the appropriate portions with the transparent ink in the colored pens in the set. Colored overlay material can also be placed on 35-mm slides.

5.6 Moving media

Although home movie projectors are still available, the videocassette recorder has become so common and easy to use that it is by far the favorite choice for producing do-it-yourself moving visual aids. All of the equipment required can be rented on a daily or monthly basis, if you or your organization does not own it. The more expensive, commercial type

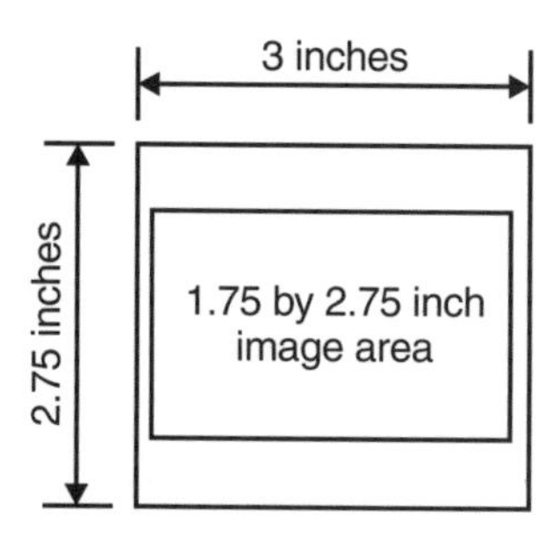

Figure 5.8 Layout dimensions for negative slides.

of moving picture cameras and related equipment can also be rented if the project warrants it.

If you have material on film that you want moved to videotape, this can be done by many commercial laboratories at a reasonable cost. These same laboratories can move material from one type of tape to another (for example, Beta format to VHS format) and will also do tape editing to your specifications, charging on an hourly basis. Many laboratories will move material from videotape to film, although this is more expensive and fewer laboratories are prepared to do it because of the much greater popularity of the videotape medium. The material in the remainder of this section is presented as though the medium were videotape, but it applies as well to film.

5.6.1 Equipment required

Depending on the type of material you want to record and the situation in which you will be recording, you will want to select either a separate video camera and recorder or a camcorder.

Camcorders, which combine video cameras and video recorders, are much more convenient to use if you will be recording in other than a studio situation. Since they are self-contained, they are more easily moved and set up than are the more bulky recorders used with conventional video cameras.

If you will be doing all of your recording in a dedicated room, the separate video camera, which is smaller, may be more convenient. Its lens will generally have a greater zoom range than the lens on a camcorder (14 to 1 versus 8 to 1), allowing you more flexibility in formatting material. In most situations, you will be able to use existing room lighting for your videotaping sessions, but if extra lighting is required, special equipment can be rented.

5.6.2 The recording process

The actual recording of the videotape is very straightforward. Use the equipment according to the manufacturer's instructions. Like anything else that you are doing for the first time, the process should be checked out well before you invest a great deal of time. Make a test run and play it back on a monitor to be sure that you have adequate light and that there are no improper reflections. In general, you should use a tripod for the video camera rather than hold it by hand. Tapes from handheld cameras are generally very hard to watch because tiny movements of the camera cause big and distracting movements on the screen when the tape is being viewed.

Although it is possible to just set up a video camera in a single location and go through the actions that you want to record, this generally makes a boring videotape. If someone who knows what to expect is running the camera during the recording session, the zoom lens on the camera should be used to allow the viewing audience to see the situation in different perspectives. This will make the tape much more interesting and easier to view.

5.6.3 Planning functions

It seems obvious, but take the time to decide what you want to accomplish in the moving medium before you start pulling tape. You may want to tape an entire briefing, or you may just want to tape an event or process so that the tape can be used as a visual aid during a conventional briefing. You may also want to tape a self-contained "mini-briefing," which forms part of a full briefing. You will need to consider how the tape will be used, who the audience will be, and the capabilities of the person who will be presenting the tape to the audience.

Taped briefing

If you will be taping a self-contained briefing (or part of a briefing), the planning functions will be just like those for any other briefing. However, the visual aids used must be compatible with the video-recording situation. You may not want to use projected visual aids (overhead projector slides or 35-mm slides) because the room lighting restrictions will make it difficult for the viewing audience to see both the visual aids and the briefer. Flip chart visual aids or material written on a blackboard or whiteboard generally make a more professional presentation on the tape.

If you must use slides, a commercial tape editor can place the material from each slide onto the tape in full-frame size for any amount of time you specify, and your voice can be put onto the tape to explain the slide during that time. However, you will not be able to point out parts of the slide while it is on the screen. A professional can also split the screen and present your visual aids on one side with you (talking and gesturing) on the other side.

Pick a recording location in which the background will look good on the tape. Having the speaker stand at a lectern before full-length drapes is one attractive option.

Consider having some kind of an audience present for the taping of the briefing so the speaker can interact with people while making the presentation. It takes quite a leap of imagination to interact with a video camera as you would with fellow human beings, so the presence of a live audience will tend to make the presentation more realistic.

It is an excellent idea to have a script for the taping. This does not need to be a word-for-word script of what everyone will say, but it should be a fairly good description of who will do what, in what order, and using what equipment.

If the briefing is fairly short, you may find it possible to shoot the whole production in a single continuous taping session. This will give you the final product without any need to edit the tape. The primary disadvantage of trying to do a tape in this way is that people will often be nervous in their efforts to say every word properly or because they dread having to start over from the beginning. The result may be an excessive number of start-overs and a very distraught briefer.

There are commercial firms that do tape editing. You can either use their editing equipment or pay them to do the complete editing job for you. In either case, you can remove unwanted material or combine taped segments into a single presentation tape. You can also do professional-looking fades from one scene to another, add background music, or perform most of the other technical functions that you see used in television programs.

Taped event or process

It is often effective to show a short film clip or segment of videotape as a visual aid during a conventional briefing. This is a good choice for any event or process in which the action is a significant part of the message you want the audience to get from the visual. Avoid throwing in a film clip or piece of videotape just for a change of pace. Like any other visual aid, to be included it should help achieve the goals you have set for the overall presentation.

To make the video segment as effective as possible, use only as much of it as is required to get the message across. The original films of tests typically show the test setup for a long time then show the actual test for a few seconds. It is best to edit this kind of tape so that your audience sees the test setup for a few seconds and then immediately sees the interesting part of the test. If the actual events shown happen in a very short time period, you can make a much more effective presentation by showing the event at full speed and then in slow motion with stop action frames at critical points in the action.

When the action takes a long time, it is best to do a speed-up of the action for the audience. The classic example of this process is showing a flower's petals unfold in a film of a few seconds' duration so that the audience can see the relative motion of the parts of the flower during the process.

Unlike fixed visual aid frames, there is a great deal of flexibility in the amount of time occupied by a moving visual medium. However, be careful to edit the film or tape so that it does not include extraneous material that will detract from the focus of your talk.

Project a computer screen live

When you are giving a briefing that includes real-time computer-generated data or computerized processes, you may want to hook the computer up to a direct computer projector and just project the screen as some process is happening. An excellent example of this was a briefing that compared various types of video compression. The computer screen had a split video display. One side was regular full-resolution video. The other side was the other half of the picture—but passed through a video compression program. The audience could watch in real time as the compression ratio was increased until the quality of the compressed video started to degrade and ultimately became unusable. The presenter would explain what compression approach he was applying (in real time) and what the compression ratio was (at that moment). It was a very powerful briefing.

5.7 Flip charts and display cards

The basic procedure for making flip charts and large display cards is to draw and write directly on the flip chart pad or card stock with broad felt-tip pens. There are several techniques that will make the end products look much more professional:

- Use light drawing guidelines for lettered text as shown in Figure 5.9. Unless they are too light to be seen by even the closest members of the audience, be sure to erase them with a very clean eraser when the text has been written.
- Remember that the size of lettering must be at least one inch for every 30 feet to the most distant audience member.
- Be careful of bleed-through from the felt pens onto later pages when making flip charts. If you skip every other page in the flip chart pad, minor bleed-through will not be a problem, and it will also be harder for the audience to see later pages through the page you are showing them. If bleed-through is very severe, use an extra

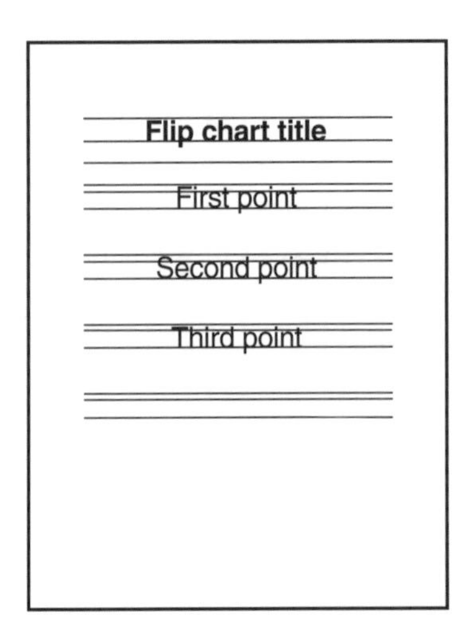

Figure 5.9 Guidelines for drawing flip chart visual.

sheet or two of paper behind the page you are making to protect lower sheets while you are drawing the charts.

Unless you are an artist, you will probably have difficulty copying drawn material large enough to be seen onto flip chart or card stock pages. The following techniques will help you avoid freehand drawing:

- Make an overhead projector slide of the drawing you want on the flip chart. Tape a blank chart page onto the wall and project the slide onto the page at any desired scale. Then you can trace the material onto the page.
- Use a metal yardstick for a straight edge when making straight lines.
- Draw circles with a felt-tip pen tied to one end of a string that is held to the page with a tack.

When using cards rather than a flip chart, you can glue enlarged photographs and computer-generated text and figures onto the card stock (using spray cement) rather than (or in addition to) drawing and writing directly on the material.

You can, of course, generate material for flip charts on your computer—using very large text. If you have a large page printer, you can make excellent chart pages, but few people have those large printers. The problem with a regular printer that only accepts letter, A4, or legal size paper is that you don't get many letters per page. The trick is to print out in landscape rather than portrait format and to minimize the page margins. That way, you may be able to get one line or at least one word on a page. Then, trim the printed material to the subject and glue the pieces to the flip chart paper.

5.8 Three-dimensional models

Just as you would do before starting to make any visual aid, be sure that you know what you want a three-dimensional model to do for you before you invest a great deal of time or money. Then, figure out the minimum requirements that the model must meet in order to meet your goals.

These include functional features, size, and total scope. Finally, determine how the model will get to the briefing room.

By gathering this information before you begin, you may save yourself a great deal of trouble. If the model is more than required to do the job, it will not only cost more to build and be harder to handle, but it will not be as effective in your briefing as a model made for your objective.

> One particularly interesting mathematics professor used to demonstrate operations on paraboloids of revolution (which look roughly like half of a football with the end pounded round) using an extremely simple visual aid. He would place one foot flat on a desk to elevate his fully bent knee. Then he would draw on his trouser leg (on and around the knee) with chalk to show what he wanted the class to understand. He looked a little strange for the rest of the day, but he was an extremely effective teacher.

There are several shortcut approaches to building the actual model that are worth considering:

- Look in school supply catalogs and stores for an existing model that is close enough to what you need to accomplish your goals. If you can buy what you want, it is almost always cheaper than building it. Also, if the store does not have exactly what you want, you will learn enough about how similar models are made to improve or simplify your design.
- Look for something that is almost what you need, then modify it to the desired configuration rather than starting from scratch. Toy stores are excellent sources for "almost right" models. One excellent example is a scale model plastic military aircraft that can be rotated over the surface of an overhead projector to show the shape of the aircraft from various aspect angles.
- If the model will only be used a few times, consider making it from art board, which can be cut with a knife or scissors, rather than the more expensive and harder-to-cut plywood that you would use for a more durable model.

- If the model is made of wood or paper, use a hot glue gun to put it together. The hot glue is strong and requires no clamping.
- Consider making the model from plastic sheet stock. It is available in many colors and thicknesses; it can be cut with a fine-tooth saw; the edges are easy to finish; it is assembled easily using special glue available from the plastic distributor; and the final model will look very professional.

5.9 Direct computer projection

Last, but far from least, you can avoid the whole problem of creating actual visual aids by just leaving your visual aids in the computer and projecting them onto a screen with a direct computer projector. This makes the use of color slides relatively easy (and cheap) and avoids a lot of messy work.

The problem is that everything must be in digital form and formatted into the briefing software format. Any information that is only available in paper form must be scanned. Normally, this is straightforward, but the file sizes sometimes get quite large and cause a noticeable delay when changing slides unless you have a very fast computer.

Transporting the slide information between computers can be fairly easy. If the files are not too large, you may be able to contain the briefing on one or two floppy disks. These are loaded into internal memory in the machine from which the presentation will be projected (to avoid the slower access time of the floppy disk). For larger files, it is practical to use one of the new high-density disk systems to avoid loading a considerable stack of floppy disks. Naturally, the projecting machine must have a compatible high-density disk drive, and there are several (totally non-compatible) types.

Another way to transport the briefing is by writing it on a programmable CD. These are accepted by most modern machines.

Finally, many organizations have local area networks that can transport large files between computers. This is great for in-house briefings, but doesn't help you if the boss really likes your briefing and tells you to take it on the road.

5.10 Tricks of the trade for creating computer graphics and slides

There are many different presentation software programs, and each is updated about once a year—so this section does not attempt to explain how to use any specific program. Rather, it covers some tricks to make the programs do what you want without requiring the highly trained hands and eyes of the professional artist.

5.10.1 Creating complex solid objects

It is often desirable to show relationships between physical objects by showing a view from some perspective of small representations of objects in the proper physical relationship. For example, you might want to show (from above) the relative positions of an aircraft, a mountain, and an airport. This is done using a graphics package such as MacDraw or the graphics in PowerPoint. It is best to create an appropriate object, so that it can be moved to the appropriate spot, duplicated, oriented, and colored as required to create your diagram.

There are several ways to create such an object. One is to draw one on paper and scan it (assuming that you can draw a little and have a scanner available). However, depending on the graphics package you are using, you may not be able to reorient the scanned figure or change its color. In most circumstances, it is better to just draw the object on the computer.

Since few but professional artists can actually make a sketch on the screen using a mouse, the following process is usually best for us common, nonartistic folks. (There is more detail on this technique in Chapter 6.)

- After opening the graphics package, set the "grid" on. This will cause all lines to start, end, and change direction at a set of points that are spaced at regular intervals on the screen (for example, one-eighth inch).
- Then, set up a framework as shown in Figure 5.10. The framework is made of rectangles and lines. If you want the object to be symmetrical, use two sets of identical rectangles—one on each side of the center line. Making one rectangle and then duplicating it is the

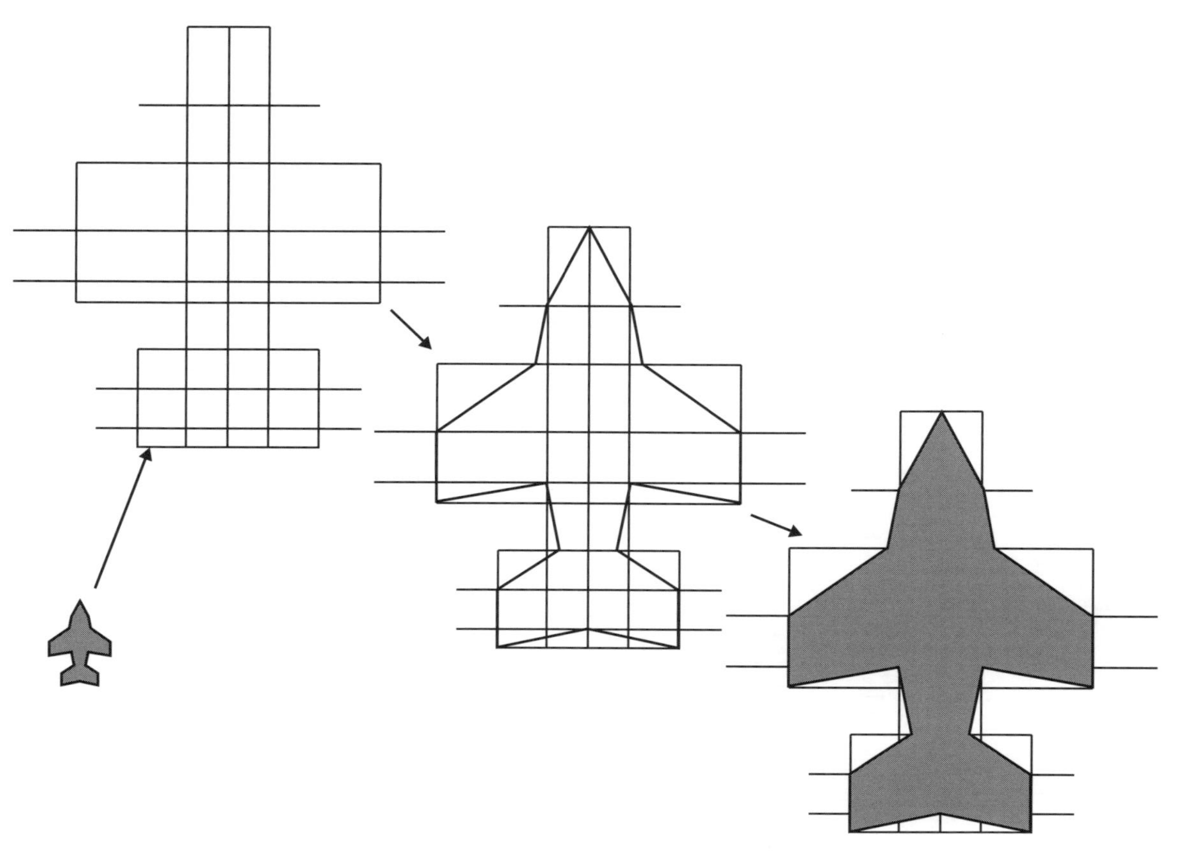

Figure 5.10 Generate the desired figure to fit an oversized layout frame.

easiest way to get two or more identical rectangles. The lines, which go across the framework, locate "corners" in the object.

- After the framework is done, you will want to save it as an object.
- Then use the "arbitrary shape" tool to draw the shape you want. Most graphics packages have such a tool. It allows you to move the mouse cursor to a point, then click to bring a continuing line from the last indicated point to the present point.
- You will notice in the second step shown in Figure 5.10 that the selected points don't need to be at line intersections. If you make them one grid increment from an intersection, you can still keep the symmetry of the object.
- Once the object is done, you can cancel the framework. If you save the framework as an object, it can be removed in a single action.
- As shown in Figure 5.11, the object is drawn large for convenience. It can be reduced in size, changed to another color, reoriented, or stretched as you desire.

5.10.2 Creating complex graphs

In a technical briefing, it is often necessary to provide complex information in graphical form. This is simplified by use of a spreadsheet program. The example in Figure 5.12 has a mathematical formula in each of the columns A, B, and C. The values calculated by these three formulas are plotted against the same variable input (which is shown in column x of the table). So far, this just follows the instructions for your spreadsheet software, but there are a few tricks of the trade to make great looking slides.

The example in Figure 5.12 is a line graph; the different lines on the graph are generally drawn at minimum width and each in a different color. To create a slide the audience can read, you will generally need to increase the line thickness. You may want to change the colors in which the lines are plotted. Some colors (like yellow) don't project well as lines.

The spreadsheet software will allow you to choose the grid line density on the chart. In general, it is better to have only a few grid lines. More

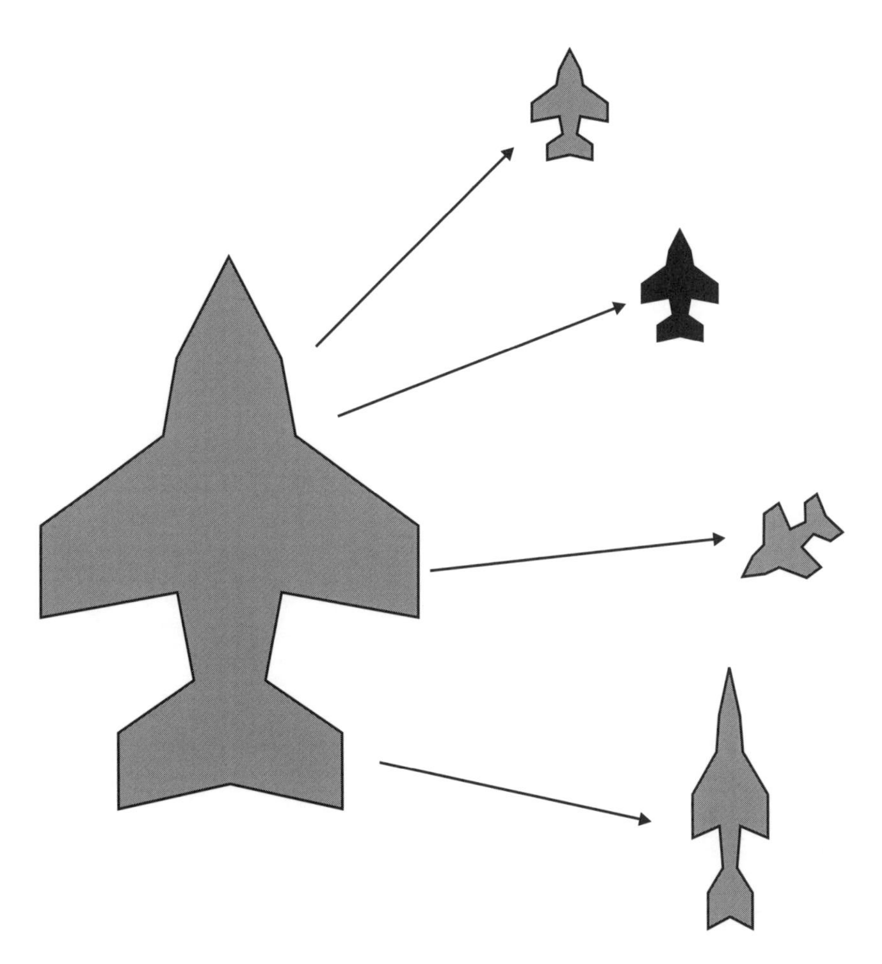

Figure 5.11 Modify the figure to your purpose.

grid lines allow you to read the chart more accurately but make it hard for the audience to see. If the audience has a handout to accompany your briefing, you may want to show more grid lines on the handout than you project on the screen.

Identify the axis lines adequately to avoid any confusion. Normally, this can be done within the spreadsheet program, but you can also add them after the graph is inserted in your slide. You can add extra information as shown in Figure 5.12 to indicate some important point you want to make with the graph.

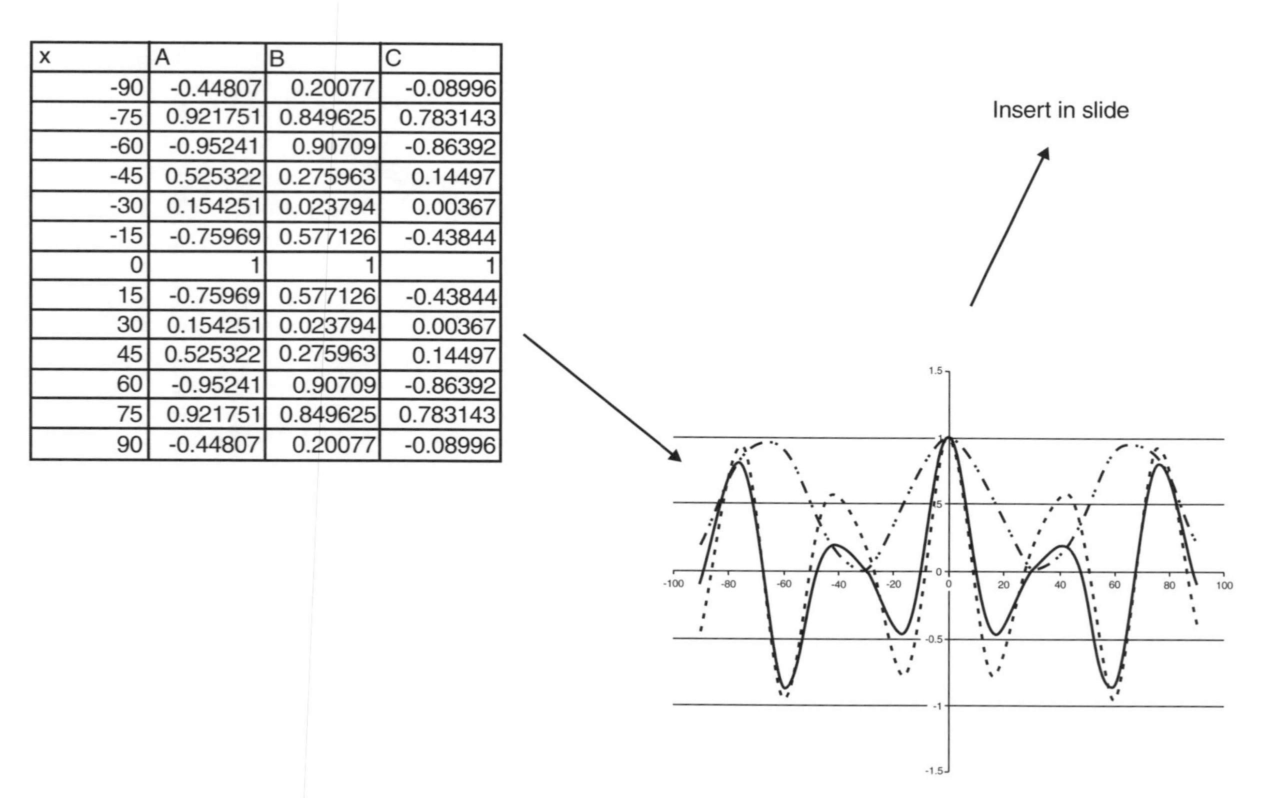

x	A	B	C
-90	-0.44807	0.20077	-0.08996
-75	0.921751	0.849625	0.783143
-60	-0.95241	0.90709	-0.86392
-45	0.525322	0.275963	0.14497
-30	0.154251	0.023794	0.00367
-15	-0.75969	0.577126	-0.43844
0	1	1	1
15	-0.75969	0.577126	-0.43844
30	0.154251	0.023794	0.00367
45	0.525322	0.275963	0.14497
60	-0.95241	0.90709	-0.86392
75	0.921751	0.849625	0.783143
90	-0.44807	0.20077	-0.08996

Figure 5.12 A complex graph generated from a spreadsheet.

The above information is emphasized because spreadsheet programs generally create line graphs in a form more appropriate to paper reports than to presentation visual aids. The other types of charts (such as bar charts and pie charts) are already well designed for presentation. The only problem is generally that some of the colors chosen by the program may not project well.

If your visual aids must be black and white, be sure to apply different patterns to areas that must be differentiated, since the different colors are represented in the black and white presentation as subtle shades of gray that are hard to distinguish. Likewise, code the different lines with different kinds of dashes.

5.10.3 Creating creative tables

One of the big advantages of using a computer to create visual aids is that you can mix many kinds of inputs. An excellent example of this is the incorporation of graphical data in a table. The easiest way to make a slide of a table is to import the whole table from a spreadsheet or a word-processing program. However, such a table will be restricted to words and numbers. Mixing in graphics can make the table much more effective as a slide.

Figure 5.13 shows several useful techniques in the generation of free-form tables. One is the spacing of lines on a table that you draw from scratch. If you make a rectangle for the outside, vertical and horizontal lines divide it into the table. For symmetry, you may want equal spacing between the horizontal lines. The easy way to do this is to make a rectangle to fit between two lines. Then, duplicate the rectangle for the other horizontal blocks and place them as shown in Figure 5.13. After the lines are in place, you delete the rectangles.

Figure 5.13 shows digital photographs in the right-hand column. They could as well be line drawings. The photos or drawings are imported into the slide as objects and sized and located. Notice that the bottom photograph overlaps the lower line of the table. This can be corrected by bringing the rectangle that forms the outside edge of the table to the front. Naturally, the rectangle must be filled with "no fill" so that the rest of the graph can be seen.

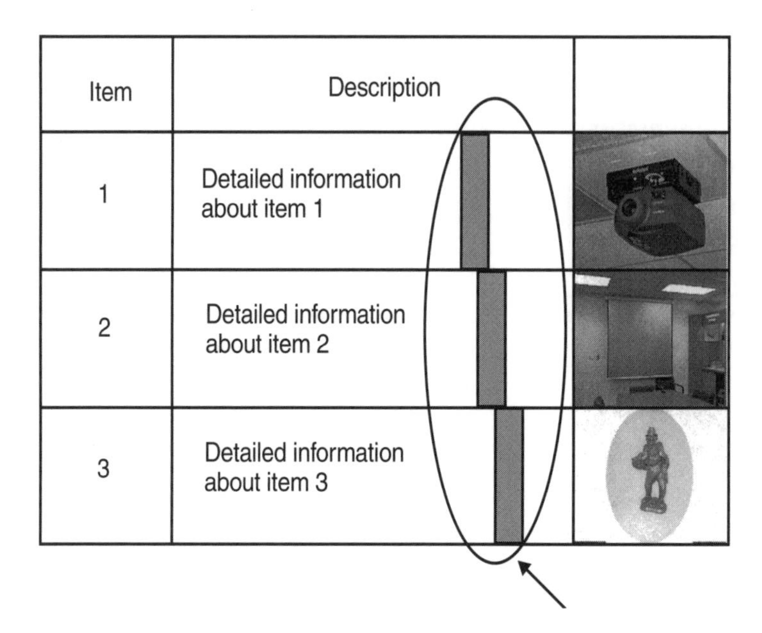

Figure 5.13 Example of a free-form table.

5.10.4 Creating full-size half-page slides

If you need to reduce the weight of a large number of slides—for example, if you are traveling halfway around the world to present a five-day course—you may want to use half-page, unmounted slides. To do this, you want two slides per page as shown in Figure 5.14. Each page is then cut in half to form two half-page slides. Whereas full-page slides include significant margin area, you will want the material on your half-page slides to go clear to the edge. Making slides in this format can be a little tricky, since presentation software is not really set up to work that way. Incidentally, the PowerPoint two-slides-per-page notes format will not work because the two slides per page will have print too small to be seen when projected.

One technique that does work is to set up the slide page format for portrait rather than the normal landscape. Then minimize the margins and enter the text and graphics in free form. The fill-in-the-blanks procedure of the presentation package is only set up to support full-page slides.

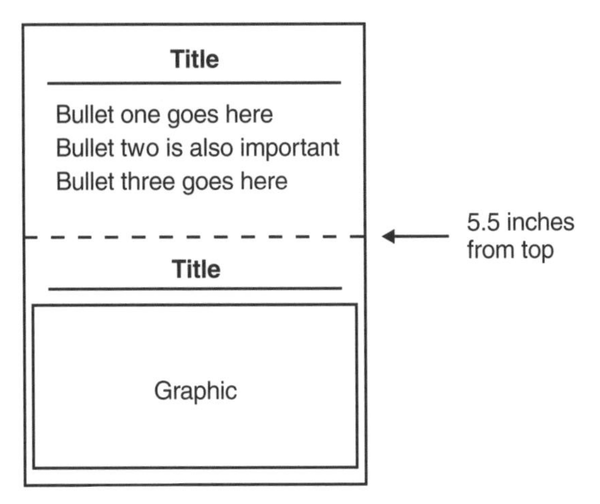

Figure 5.14 Two full-size slides on one page.

Another technique is to print out full-page slides (on paper), crop them to the material, and mount them on a piece of card stock slightly larger than a regular piece of paper. Then, make the actual slides on a copy machine, which will reduce the oversized pasteup so that each of your slides fits neatly onto a half page.

6

Getting the Most Out of the New Presentation Technology

There is an impressive array of new computer hardware and software available for the briefer, bringing three new and rather compelling concerns to the presenter. The first concern is the decision to use none, some, or all of the items; the second is a whole range of new ways to get into trouble with them; and the third is how to keep existing material usable.

The good news is that the new equipment expands your range of options. Color presentations are now easy and fairly inexpensive (once you have the equipment). You can mix fixed slides with video clips, even on the same frame, or display actual computer interactions in real time for supereffective demonstrations.

You can make slides with professional-looking text and graphics using clip art for added impact or add all kinds of eye-catching transitions between slides and build the material on a single slide using standard software features.

Because this new technology changes so rapidly, this discussion will be limited to the general types of services and software available, the trends in development of new devices, tips for staying out of trouble while using this new technology, and advice for getting the most out of this family of new hardware and software.

A series of software programs to help presenters generate briefing slides has been marketed over the last 20 years. Some are simple graphics programs that can be applied to briefing slides, and some are designed specifically for the presenter. As this goes to print, there are about 10 such programs being widely used in the technical community. Many of these are upgrades of previous versions with additional features or new ways to transfer information among related word-processing and spreadsheet programs belonging to the same "family." At any given time, there are usually three or four in wide use.

All of the new hardware and software items come with manuals instructing you on the use of a particular item and providing you with the program-specific vocabulary to know how to perform functions such as selecting commands. However, there are some important things that manuals do not tell you, and there is a body of wisdom that will only come from using these tools over an extended period. This "wisdom" comprises little tricks to get the computer to do what you want it to do and to maximize the beneficial impact of the technology on your briefing effectiveness. This section deals more with this acquired experience than with the program specifics.

Generally, the more a computer program does for you, the more it can do to you. All of the computer word-processing and slide-preparation programs make assumptions about what you are trying to do, and the computer helps you do what it thought you wanted to do. When the computer guesses correctly, this is a great help; when it guesses incorrectly, it can be infuriating. For example, try typing dBm (a type of unit used in radio communication); unless you have changed the default, the computer will change it to Dbm every time, assuming that you have typed it incorrectly.

6.1 When the media interferes with the message

There is a danger in allowing a too elaborate presentation to interfere with your message. Usually, when people see black type on a white screen, they read it, but when you use color, people may see only the color. Color can cause people to focus on one part of the information on the screen, and too much color may diminish their ability to process the needed information.

Try this experiment: Write a sentence in large, bold type, and change each letter to a different color. Now try to read it. This is, of course, an exaggeration, but the point is that if there are too many colors on a chart, the audience becomes confused. Likewise, if there are too many colored objects, the audience will notice the colors rather than the message you want to convey.

You can also generate slides in which items zip in from any side or appear one character at a time, or you can add clip art to place little pictures around the information. When used in moderation and restricted to "punch line" roles, these special effects are extremely effective, highlighting important points to make them memorable. When used to excess, they cause audience overload. To summarize, too many fancy computer touches can kill the effectiveness of the briefing—use them with moderation and purpose.

6.1.1 Predictability

PowerPoint 97 offers a number of intriguing formats that can be used for slides. They have great colors and add drama to slides that might otherwise consist only of dull data. However, at this time, there are only a few of these formats available—and everyone uses them. Thus, when you use one of those formats, many members of the audience will have seen your fancy format before, with the unintended consequence that they will know that you are using the easy way to look sophisticated. In the past, these formats were generated by an art department at great cost; audiences were surprised; and the formats added uniqueness to your presentation, increasing audience interest.

When you use a standard PowerPoint fancy format, the surprise is gone. This is a rather subtle point but an important one. It is good to have yourself thought of as sophisticated. It is *not* good to have yourself thought of as trying to look sophisticated. At best, the audience will be impressed with the format, while under normal circumstances, the audience will not even notice the format because it will be familiar. At worst, however, the formats will obscure some of the material on your slides, reducing their readability. Worse yet, the audience may believe that you are adding artificial fanciness to cover a lack of content.

6.2 Choosing equipment for computer-aided briefings

The equipment listed here focuses mainly on direct computer projectors. The other two equipment choices that affect the computer presentation are the computer that drives the projector and the storage medium that holds the briefing material. For completeness, we will also discuss two different types of computer-driven presentation media: the PC-interactive whiteboard and integrated video presenters.

The most important quality that most briefers want in their equipment is standardization: They want the equipment to work together, and they want to be able to transport their briefing material to other locations or other facilities and be able to present their material without equipment incompatibilities.

Fortunately, projectors are reasonably standard, in that most work with most computers. However, the standard for briefing storage media is more transitional and thus takes more coordination. The main computer-related compatibility challenge stems from the differences between Macintosh and PC formats. Software can be transferred from one to the other, but the transitions are less than trouble-free for the average, unsophisticated computer user. There is also a compatibility problem with the software versions used, which is discussed relative to updating a previously presented briefing in Section 6.5.

6.2.1 Projectors

Currently, the best technology available for computer briefing is the direct computer projector, a device that emulates the computer screen. The first such projectors were large three-gun projectors suitable for large, fixed facilities, and their lack of portability limited their acceptance and usefulness.

The first portable projection devices fit onto an overhead projector. They were, in effect, computer-driven overhead projector slides. Anything on the computer screen could now be projected onto a screen for viewing by an audience of moderate size. These devices met with indifference because the images were weak. However, direct VGA projectors overcame this shortcoming and were an instant success.

VGA projectors accept signals from the VGA connector on almost any computer and project whatever is on the computer screen. They are now available in a wide range of sizes—size, weight, and power draw being proportional to the room size served. Most VGA projectors are equipped with remote control devices to allow an instructor to change slides, focus the projector, and move a pointer around the screen without direct access to the computer.

The smallest VGA projectors weigh less than five pounds and are easily carried in a briefcase. Small projectors in this weight class are available from several manufacturers but are still quite expensive.

Table 6.1 shows the range of direct computer projectors now available. Since new devices are brought to market frequently, surveys of new products appear in computer magazines on a regular basis. Figure 6.1 shows a number of modern VGA projectors.

6.2.2 PC-peripheral and -interactive whiteboards

Whiteboards that allow their contents to be captured have been around for several years but have recently become more useful with a number of PC-interactive models with different features. The basic product is a large whiteboard on which you write with regular dry markers. The old style boards have a vertical bar that sweeps across the board on command. As the bar moves, a printed copy (black-and-white) of the contents of the board comes out of a thermal printer mounted to the bottom of the board. The idea was to record the notes from a meeting. Copies were (and are) typically made for all meeting participants. These devices were very popular in large organizations, and many are still in service.

Today, there are two main types on the market (see Figure 6.2). Both are interactive with computers and offer many additional features.

One type is called a PC-peripheral board. It allows you to capture the board contents, print copies, send copies over the Internet, or

Table 6.1
A Selection of VGA Projectors

Application	Brightness (Lumens)	Resolution	Weight (Pounds)	Price
Large-room fixed facility	1,000–1,600	800 x 600	29–30.5	$9,000–10,000
		1,024 x 768	29–30.5	$10,500–12,000
		1,280 x 1,024	29–84	$22,000–26,000
	2,000–3,200	1,024 x 768	50	$21,500
		1,280 x 1,024	39–76	$24,000–30,000
	5,000–7,000	1,024 x 768	100–238	$60,000–120,000
		1,280 x 1,024	200–207	$85,000–95,000
	9,500–12,000	1,024 x 768	225–250	$90,000–115,000
		1,280 x 1,024	180–250	$110,000–150,000
Portable	500–800	800 x 600	10.3–25	$3,800–7,000
		1,024 x 768	11–18	$5,000–8,000
	850–1,200	800 x 600	14–17	$4,300–7,800
		1,024 x 768	10.2–17	$7,500–9,600
		1,280 x 1,024	16.1	$9,600
		2,048 x 1,536	21.9	$19,000
	1,300–2,200	800 x 600	13–20	$7,300–9,000
		1,024 x 768	13–20	$8,000–16,000
		1,280 x 1,024	16.8	$13,000
Ultraportable	280–330	800 x 600	9–9.9	$2,900–3,000
	500	600 x 600	9.9	$2,700
	500–800	800 x 600	5–10	$3,000–7,500
		1,024 x 768	4.2–10	$4,500–9,000
	900–1,100	800 x 600	7.4–10	$4,000–6,000
		1,024 x 768	7.4–10	$5,000–10,000

incorporate board data into slides in a briefing. The PC-peripheral boards cost $500–600.

The second type, which is called a PC-interactive board, works with an LCD projector. The interactive board allows you to project material from the computer and then interact with it. Some have touch-sensitive surfaces that detect the marker on the board and capture the image on the computer screen. This allows the whiteboard to function as a giant

Figure 6.1 Modern VGA projectors.

touch-sensitive computer screen. PC-interactive boards cost approximately $2,000.

6.2.3 Video presenters

For permanent classroom situations, there is a new family of products known as video presenters, which are basically video cameras that input data to computers for projection. Some include computers, and some interface with external computers. See Figure 6.3.

With a video presenter, you can lay paper or objects on a flat surface. Unlike the outdated opaque projectors, video presenters do not require hot lights, which would occasionally ignite paper, or a loud fan for cooling. They allow a professor to lecture using visuals from an actual textbook. The surface is lighted from above, and the image is projected. Some have secondary cameras that show the instructor's head and body for projection onto the screen. As video presenters produce pictures in the

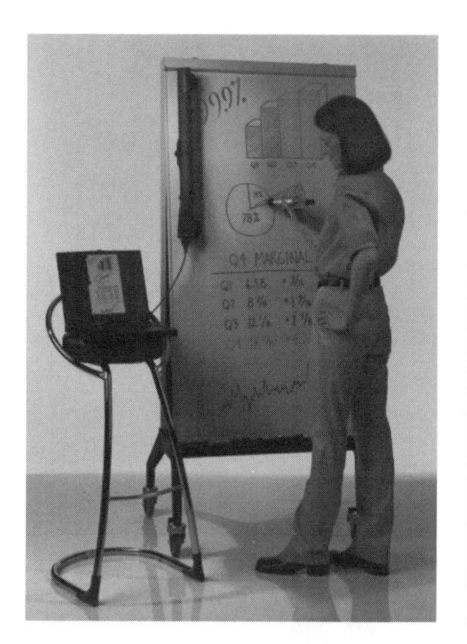

Figure 6.2 Here are two of the many whiteboard products available on the market. The SMART board is an example of a computer interactive board. The Virtual Link system captures drawings to computer files using special pens on plain paper. (Courtesy of SMART Technologies and Virtual Link Corporation.)

formats of digital cameras, their images are "pictures" rather than objects. Thus, they can be sized and cropped but not modified. Table 6.2 compares the features of two typical video presenters.

6.2.4 Computers to drive video projectors

The projector needs to be driven by a computer with a VGA connector. In a classroom or other fixed briefing facility, it is common to have a dedicated desktop computer in the room. It is also common to connect that computer to a local area network (LAN), so that individuals throughout the organization can prepare briefings on the computers at their own desks. Briefings are then transferred to the briefing room computer over the LAN. Frequently, there is also an ability to load directly onto the briefing room computer from other software media such as a floppy disk or Zip drive.

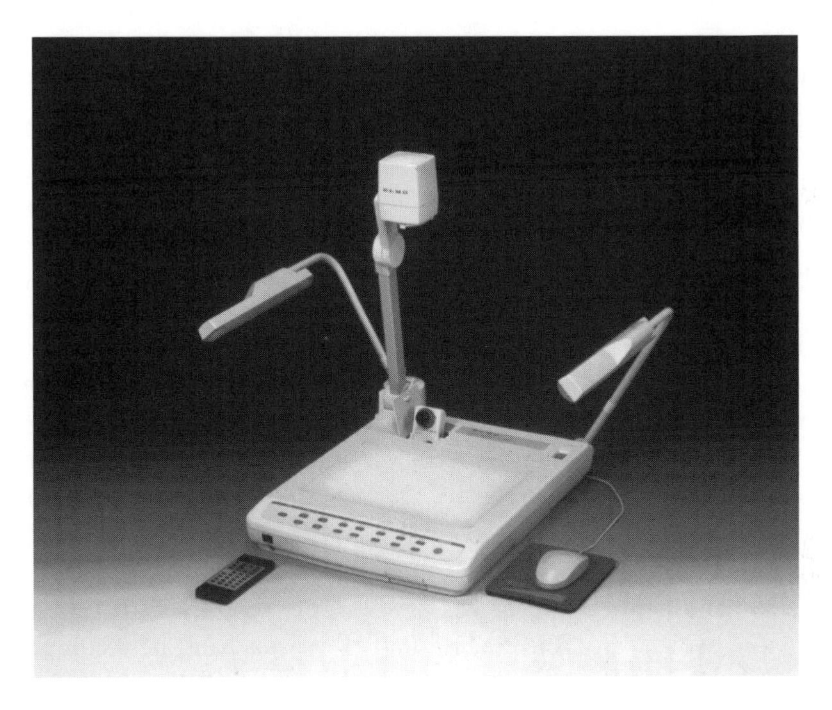

Figure 6.3 This is an example of a modern presenter product. (Courtesy of the ELMO Manufacturing Corporation.)

Table 6.2

Comparison of Typical Video Presenters

Model	Features	Price
ELMO EV-6000AF	Two video cameras (one for documents and one for 3-dimensional perspectives or instructor "head shots"); JPEG image storage and transfer; RS-232 remote-control port; mouse or remote control	$4,495
SAMSUNG SVP-6000	One video camera; JPEG or bit-map storage and transfer; RS232 and USB ports; mouse control; internal computer	$5,700

For portable applications, the most commonly used computer is a high-capability laptop. The laptop typically weighs on the order of only 10 pounds but contains all of the relevant capability of a desktop computer. Some laptop computers have internal high-density storage media, and others can be augmented with portable external drives. From the laptop computer, the instructor can show the slides and modify them in real time.

The "hand held," a new type of smaller computer, offers several advantages over the laptop computer: First, it is significantly smaller and lighter. In addition, it runs for about 15 hours between battery recharges (versus a little over two hours for the typical laptop).

Hand-held computers use special versions of software to reduce their memory requirement. Currently, they use the Windows CE operating system and CE versions of the Windows office suite programs—including PowerPoint CE.

The hand-held computer has a VGA port to connect it to a VGA projector, which can project a briefing. However, hand-held computers offer only a limited capability to edit the briefings. Only the text on the title slide can be modified—to customize the briefing to a specific audience. The body of the briefing, including all graphics, must be projected as stored.

Another inconvenience of the hand-held computer is that it does not accept data from floppy disks or Zip disks. It accepts its data via a serial cable from a laptop or desktop computer. This means that all development and even the tiniest edits must be made at home on your desktop or full-function laptop computer. The whole modified briefing is then loaded onto the hand held for presentation in a relatively unchanged format. Because hand-held computers are new and very popular, look for upgrades to overcome all of these inconveniences.

6.2.5 Storage media

Once developed, a set of briefing charts needs to be stored in some sort of computer memory, such as on the hard disk of a computer. If the briefing is contained in a large file (with a lot of high-density graphics), keeping it on the same computer is particularly convenient. The computer on which the briefing is developed also offers the advantage of software compatibility. If the development machine is a high-capability laptop computer, it can be taken to the briefing site anywhere in the world and be used to present the material.

Unfortunately, however, it is usually necessary to transfer a briefing to another computer, either to share the information with someone else, or because a different computer is available in the presentation room. It is also highly desirable to have some way to permanently store the briefing material away from the host computer, freeing up disk space on the development computer, while providing a backup in case the computer crashes.

Storage devices are constantly being improved—providing faster access and increased storage capacity. As a result, there is no absolute standard that one can confidently assume to be available everywhere.

The closest to a standard storage medium is the high-density floppy disk formatted for PC. The main drawback is that it holds only 1.44 MB of data. While a briefing with only word charts might fit on a single floppy, any significant use of graphics will require multiple floppies. This requires that the briefing be divided into floppy-size segments, which is sometimes extremely inconvenient. High-density Macintosh-formatted disks hold slightly more data but still require that any briefing with significant graphics be segmented.

To avoid having to segment briefings on multiple disks, a presenter can use a high-density removable disk system. The first such system available was the Syquest drive. Syquest disks hold 44, 88, or 105 MB of data. The problem is that relatively few computers have Syquest drives. At one point, they were widely accepted in the artistic community—with one color painting often filling an entire 44- or 88-MB disk. However, they did not catch on among the groups that do most of the computer-assisted briefings.

The most commonly used high-density removable disk system at the moment is the Zip drive. Zip disks hold 100 MB of data—enough for most briefings, even if they include color pictures and other high data-density features. The 100-MHz Zip drive with PC formatting is as close to a universal storage/transfer medium as we are likely to have. Perhaps a quarter of the computers you are likely to be using to develop and present briefings have Zip drives. This varies widely, since some organizations have Zip drives on every computer and others have never heard of them.

There are three different versions of the Zip drive. The earliest has a parallel interface to the computer. This is the slowest interface, and you must turn off the computer before connecting it to avoid damaging your computer motherboard. Next came the SCSI interface Zip drive, which is faster but may lose data if anything else is on your SCSI port; this is a major inconvenience because SCSI ports are usually used for many interface devices. The newest Zip drive uses the USB interface. This is the fastest interface and can be disconnected from your running computer without damaging either the computer or the drive.

However, there are several new storage devices on the horizon. Removable gigabyte disks and a whole array of solid-state mass storage devices are currently available, and one or more is likely to replace the Zip disk as the standard within a year or two.

The best advice is to stay loose on storage media. Be sure to transfer any old files that you value onto a current medium before your organization decides get rid of the old mass storage hardware (which some organizations do every year or so).

6.3 Using the technology to maximum effect

This is a good time to remember that no fancy equipment can make a bad briefing into a good one. The briefing has to be properly planned and timed and then the computer features can be added to give it extra interest and impact. The following sections present some examples of ways in which computer tools have been used effectively to enhance briefings.

Example 1

In this example, the briefing is part of a class for operators of a piece of computer-controlled equipment. All of the operator controls and indicators are handled at a computer terminal. The briefing starts with an overview of the system block diagrams and bullet charts to explain points about system configuration and function. Then, the operator functions are defined using bullet charts and inserted photographs of an operator using the actual equipment. There are also picture slides showing screen displays with added arrows and drawn boxes to go along with the word descriptions of the operator functions. All of this data has been presented using direct computer projection of PowerPoint slides.

Finally, the projector is connected to a computer programmed with the actual operating system from the equipment the operators are being taught to use. The software has been slightly modified to allow simulations of various situations the operators will be required to handle. The student operators see the actual screen displays they will be using. The instructor performs the proper operating functions while explaining them to the students, who see the results on the projected computer screen. When students ask questions about various operations, the instructor can perform the operations to provide real-time feedback to the students.

Example 2

In this example, the briefing covers a range test of a system that is supposed to protect an aircraft from attack by a missile. Direct computer projection is used. First, there is a series of bullet charts to explain the specifications to which the system is tested and the test set up. Graphic charts are used to show the testing schedule and the series of events in the range testing. Then, slides, including inset video clips of high-speed camera data, show the terminal few milliseconds of the engagement. Some clips show the missile hitting the target drone (when the system is not applied), and some show it missing the target drone (when the system works properly).

Example 3

In this example, the briefing aims to teach maintenance personnel how to remove equipment items from an aircraft for repair. The briefing starts with bullet charts of the maintenance steps. Then it has line drawings of the location of the equipment on the aircraft. Subsequently, a series of photographs taken with a digital camera show the hands of an expert actually removing and replacing the equipment from the aircraft. Finally, bullet charts are used to explain safety procedures.

Example 4

In this example, PowerPoint is used to prepare briefing slides. A handout is also generated, showing each slide reduced by 50% at the top of the page and a series of bullets on the bottom half of the handout page. The bullets summarize the comments the briefer makes while each slide is on the screen. The handout is printed in the PowerPoint notes view, and the slides are projected from the same data file (in slide show mode).

Example 5

In this example, a new briefing is to be assembled from a number of previously generated briefings. Each of the previous briefings has some elaborate slides that would take significant time to generate. A new PowerPoint file is opened with a title slide and a few subject divider slides to block out the major segments of the new briefing. The old briefings have been loaded onto the same computer or are available on Zip disks. With the new briefing open on the computer, the developer reviews each of the previous briefings in turn, using the PowerPoint "import slides from file" command. He or she selects the appropriate slides from each file importing them into the new briefing. Then, the "slide sorter view" is used to move the imported slides into the appropriate order and to identify any additional slides that must be generated for continuity and completeness.

6.4 Issues concerning traveling with computer slides

There is a great deal of fear associated with taking a computer-projected briefing on the road—particularly to a foreign country. The fear is that you will get to the distant, exotic land to give this most important briefing, and the equipment won't work. This is the stuff of nightmares! Fortunately, however, there are several approaches that experienced computer briefers use to minimize the chances of this happening.

Approach 1: Have a set of backup slides

Many experienced briefers simply will not travel with only a computer-projected briefing. The computer might crash, or there might be some kind of equipment incompatibility. Pure and simple, the briefer has a full set of slides on plastic in his or her briefcase. Because you can always get an overhead projector, those slides might save your briefing. Slides may not have the flash of the computer briefing, but the show will go on.

Approach 2: Take all of your own equipment

Many experienced briefers are simply unwilling to take the chance that some part of the equipment at the remote briefing location will not be compatible. Accordingly, they travel with their own projector, laptop computer, cables, spare projector bulb, and power and plug adapters. This can mean carrying a lot of hand luggage, since you typically don't want to check any of your briefing material or equipment as luggage—it might get lost or broken from rough handling. In addition to preventing compatibility problems, this approach has the advantage that, since briefers know how to use their own equipment, they will not have to learn how to use new equipment right before a presentation. The popularity of this approach explains why the industry is developing projectors under five pounds.

Approach 3: Go fearlessly where others fear to tread

Some briefers, even some who are quite experienced, refuse to be paranoid about what might go wrong: They coordinate ahead of time to be sure that the right equipment and storage medium are available at the briefing site. Then, they just slip a Zip disk into the briefcase and get on the plane. The good news is that this will work most of the time, and as computer projection equipment becomes more widespread throughout the world, it will be even safer. In fact, we will probably see a time when computer projectors are as universally available as overhead projectors are now.

Approach 4: Modified go fearlessly approach

This is a modification to approach 3. Yes, the briefer gets on the plane to the other side of the world with just an electronic copy of the briefing. The difference is that he or she has sent a sample of each type of slide ahead as e-mail attachments—well in advance of the briefing date. A local colleague at the briefing site tries the material on the site's equipment.

An even smarter tactic is to send a preliminary version of the whole briefing ahead on the exact same medium that will hold the final briefing. When the briefer hears the soothing news that the early version works on the equipment and software that will be used, well-justified fearlessness takes over.

A few hints about remote computer projection briefings

In general, if you take your briefing to a remote site on some kind of portable medium (such as a floppy disk or Zip disk), it is a good idea to transfer the briefing onto the hard disk of the computer that will be driving the projector. This makes the briefing go much more smoothly because the computer reads data more quickly from its own hard disk than it does from floppy or Zip disks. The slides, particularly those with large graphic files, will appear on the screen without the extensive delays often required to read them from storage media.

If you have a small printer with your laptop computer and want to print out overhead projector slides, be careful that your slide material is compatible with the printer. These small printers are ink-jet types, and much of the slide material available on the market is designed for use in laser printers. If you print slides on laser material with an ink-jet printer, the ink will not dry and will likely smear.

If you are planning to project a videotape through the VGA projector, be careful about the recording format and basic TV screen format. The frame rate, line rate, and synchronization of TV signals in the United States, Europe, and Australia are all different—extremely different. A tape simply will not run on a system different from that in which it was recorded. In general, it is best to have the tape professionally converted before you leave home and to acquire a projector locally.

Conference centers often have very complex video players that can play tapes of any format, but, frequently, these do not yield good results. The audio and video may be distorted, or (and I've actually experienced this) the speed of the video may be off.

6.5 Updating an earlier briefing

One of the problems associated with computer-assisted briefing is that the data must be compatible with the computer software—whether it is PC- or Macintosh-based. Moreover, you need to have the right version of the software (which the software company will upgrade about once a year).

In general, you can open a file created in an earlier version of a specific program using a later version. For example, a PowerPoint 4 file can be opened on a machine that is using PowerPoint 97. However, the machine will automatically covert the old file to the new software version. See Figure 6.4.

In general, you cannot open a file created in a later version on a machine that is running the earlier version of the same software. For example, you cannot open a PowerPoint 97 file on a machine running PowerPoint 4.

Consider this very common briefing preparation problem: You have an existing briefing whose material you haven't used in a couple of years, but you worked hard on it and have some really neat—still applicable—slides. You need to make a new briefing that is about 10% changed from the former one. The following are a few realities and techniques that can help:

- If you are updating to the next generation of a program on the same machine, you will probably have little or no difficulty. Software is generally "forward-compatible," meaning that the later version of the software will read the earlier version. However, it will convert it to the new version unless you take specific action to prevent this from happening. See Figure 6.5.
- New software versions generally carry basic keystrokes forward but sometimes are creative in the conversion of formats and graphics. You may need to do some significant reformatting—at least

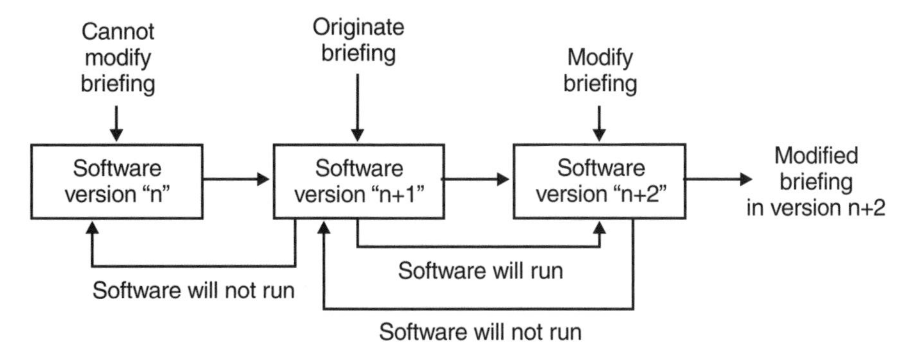

Figure 6.4 Moving between software versions.

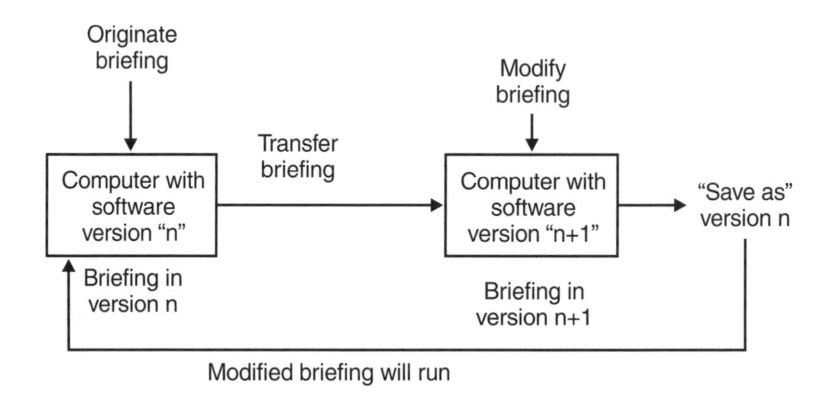

Figure 6.5 Saving in the old software version.

check the formatting very carefully after performing any software conversion of briefing slides (or any other material for that matter).

- You generally cannot use an earlier version of software to modify a presentation developed with a later version of the same software. This means that when you have modified your presentation on someone else's computer (with the new version of the software), you will not be able to work on the modified briefing on your old software. The way to solve this problem is to save the modified briefing in the earlier version. (Most programs will let you do this.)
- (This is a real-world example.) If you have a presentation that you developed in MacDraw 1.6 on a Macintosh Plus machine (using the operating system that came with it) and you try to load it onto a Macintosh Performa (using the operating system that came with it), the new machine will crash. Not only won't it read the software, it will likely give you little choice but to shut down the machine.

Fortunately, if something like this happens to you, you have a few options:

- Find an old computer that will run your software and make your modifications on it. This is very tempting, but it only solves your short-term problem. Sooner or later, you will have to modify it again, and the older computers may not be available.

- Find that older machine, and convert the old program to a newer program that will run on the new machine. Then, on the new machine, convert the intermediate program to the new software and go right on with your modification. This has the advantage that you now have a version that is modifiable the next time. This is about the best solution if you have that old machine and appropriate software available.
- Scan hard copy of your old slides and place the old slide information onto slides in the new program on your new computer. This is also tempting, but it has some similar drawbacks: If the material contains graphics, the scanner just makes a picture of the old slide—it doesn't recreate the slide. This means that you cannot modify anything on the scanned slide. Furthermore, if the hard copy you are scanning is a copy of a copy, it will probably have annoying black spots between objects or letters, as the scanner will faithfully copy all of the background clutter along with the information. Finally, scanning sometimes creates files that are very large. This means that one slide with a scanned image may take more memory than several newly created slides. Large files often take a long time to come up on the screen when you change slides. If the data is text, the scanner can capture the keystrokes, but the formatting is seldom satisfactory.
- The last choice is to look at a hard copy of the old material and create it from scratch in the new software on the new computer. This sounds like a fate worse than death but is sometimes a very reasonable alternative for several reasons: First, you may wind up taking more time fiddling around trying to convert the old material than it would take to do it over. It is a great temptation (and common occurrence) to spend three days trying to save something that could be redone in a half day. Second, it is generally much faster to copy something than to think it up the first time. Third (and most compelling), if you do it over, you will probably make it better.

It is important to note that you can generally make a copy of the presentation in an earlier version of the preparation software. In

PowerPoint 97, you use the "save as" function, then select the earlier version. Use caution, however: Sometimes the files become huge (using 10 times the memory) when converted to the old program.

6.6 Classified information

If you must give a briefing that includes classified government information, the computer briefing brings an additional complication. Classified information comes with protection requirements that are sometimes more "theological" than technical in nature. Any time that classified information (even one bit!) is stored in a computer, the user is required to certify that there is no possibility that any classified information remains in the machine or to protect the machine to the required level (i.e., lock it in an approved safe).

As you may know, when you delete a file from a computer, you actually just remove its address and allow other information to be overwritten in that part of the computer memory. The actual file data is still in the computer. In fact, much of the information from the file could be stored in several temporary files that the computer software automatically generates. Thus, an expert with the proper equipment could recover the data from the computer.

If 100% of the written memory in a computer is volatile (i.e., loses all information when power is removed), there is no problem. Just turn off the computer after the classified material is removed, and the computer can be used for any other purpose. The problem is that almost all computers now have internal hard drives or other nonvolatile, writable memory. If any classified information has been in a computer with nonvolatile, writable memory, the whole memory must be wiped clean. That includes the operating system and all software. There are special programs that write a one and then a zero in each memory location—then write a random bit pattern into all memory. Before the computer can be used for any purpose, it must be reloaded with the factory software configuration and, subsequently, with all of the software you intend to use.

The best way to avoid this problem is to be careful not to put any classified information on a general-use computer. Computers can be configured with removable memory so that the entire nonvolatile memory module can be removed for secure storage. Such computers are often provided for classified briefing applications.

Incidentally, VGA projectors have internal memory. However, this memory is volatile, so cycling power to the projector removes any classified information.

6.7 Graphics issues

There are two ways to add graphics to a briefing slide. One way is to create a drawing, table, or chart. The other is to import it as a picture. The primary difference is that the input that you create on a computer can be modified. You can edit it to add or subtract information, or you can modify the color or other features at will. On the other hand, a picture that you import cannot be modified. It can be cropped or sized, and you can overlay information onto the picture, but you cannot modify the picture itself. In most cases, you cannot even rotate it in the drawing.

First, we will consider how to get pictures into the briefing. This can be done by scanning or by using a digital camera.

6.7.1 Scanning

A wide variety of scanners are available. Some are extremely portable, while others are only useful in an office environment. All create digital files of printed information. The primary usefulness of a scanner in preparation of a briefing is to create digital files from drawings or photographs. Scanned photographs can be color or black-and-white. Most scanners come with photo-processing software that will allow you to create all sorts of artistic effects with scanned photographs.

Scanners have various levels of resolution. The least dense scanners are directly compatible with the density on a computer screen—which is the level of density that is projected from a VGA projector. In general, if you scan with more density, it just increases the size of the file—without improving the quality of the projected image.

Scanner software will allow the generation of several types of files. It is important that the file type be compatible with the presentation development software you are using. For example, JPEG files are easily inserted as "pictures" into PowerPoint 97 slides.

Scanned images can be cropped and sized for incorporation as pictures into briefing slides. Because they are pictures (rather than objects), they cannot be modified in the briefing preparation program.

6.7.2 Digital cameras

The other way to create digital photo files is by using a digital camera. There are many digital cameras on the market, falling into three categories: One type has very good resolution, using internal storage. You need to hook this type of camera up to the computer in order to read it out. The second type also provides high resolution but uses special cartridges or disks to record the images. These cartridges can be quite expensive and must be read into the computer using either the camera itself or a special adapter. The third type of digital camera currently available puts the images on standard IBM-formatted disks that are inserted into the camera. This type, however, provides lower resolution.

The cameras that store the images internally have the disadvantage of limited storage. You can take only a limited number of photos before having access to a computer. The type with special memory modules can be supplemented with extra modules to increase the number of photos while you are out of contact with the computer. However, these modules are expensive and can typically be purchased only where the camera is sold. Moreover, you need the camera or the special adapter to upload the picture files to the computer, so it is normally not practical to mail the disks to someone else for inclusion into a common briefing.

The direct floppy disk cameras have the great advantage of putting the pictures right onto standard high-density floppy disks. The disks are cheap, and you can get them almost anywhere. You just plug the disks into the drive on the computer and transfer the files as required. The disadvantage of this type of camera is that it has lower resolution, which can sometimes be a problem. For example, the resolution of pictures from a disk camera was not high enough to permit their use in this book.

However, slides do not require extremely high-resolution photographs. The disk camera pictures have sufficient resolution for projection by a VGA projector.

All of these cameras take special (rechargeable) batteries, which can be expensive. If you buy a high-quality, long-life battery, it may be less of a problem. You should be able to take plenty of pictures and then recharge the battery overnight. However, if you are going somewhere lacking readily available power and/or need to take a very large number of pictures in a single session, you may find an extra battery a necessary investment.

6.7.3 About high-resolution pictures in slides

It is easy to get too much resolution in photographs that are used in slides. Most scanners are capable of providing several times the resolution that is projected. With too much resolution, file size may grow too large (since greater resolution causes greater file size). Doubling the resolution causes the file size to increase by a factor of four. Even though the amount of memory in computers is huge and increasing almost daily, there are still two very practical problems associated with large files: transportation and projection speed.

A floppy disk has only 1.4 MB of capacity. The digital file for a single high-resolution photograph can easily exceed that. This means that you cannot move the file by floppy disk—which is sometimes the only way that is conveniently available. A standard Zip disk holds 100 MB, but if your briefing has many high resolution photos, you can even exceed that. If you are sending photos over the Internet, your modem will have some maximum transfer speed, so it takes more time to transfer larger files. It is a great inconvenience to wait a half hour or more for a large file to arrive.

When you are presenting your slides, you give a computer command that causes the next slide to be shown. Then, you must wait until the computer puts the slide onto the screen. The larger the file, the longer this takes. Modern computers are so fast that you usually don't notice the delay. However, if the next slide has a very large file (for example, a high-resolution photograph), it can take several seconds for that slide to come up. This can break the rhythm of your presentation.

6.8 Tricks for the novice computer artist

The manuals that come with graphics software programs are very valuable in that they tell you how to access the commands and features of the program. However, there are a few things they do not tell you about how to make professional-looking graphics.

As previously mentioned, all of the graphics programs have about the same features (based on the author's experience with 10 different programs and versions on both Macintosh and PC). Here are a few tricks that will help you make professional-looking graphics on any of the programs. PowerPoint 97 is used as an example to illustrate techniques (because it is currently on the author's computer), but any of the other programs would work as well.

Use clip art sparingly

It is a great temptation to use clip art in graphics slides. The clip art is professionally drawn and looks great. However, it looks like clip art: It is never quite the right picture and will be stylistically different from the rest of what you want to do in the graphic. Thus, it is recommended that you use clip art sparingly and for special effect unless it is specifically what you need to tell the story you want to tell.

Use the grid

All graphics programs have a mode in which all objects are aligned to a grid on one-eighth-inch or one-tenth-inch centers. By using the grid to draw block diagrams or other figures, you can assure that your lines meet each other exactly. Arrowheads go right to the edge of the box. Arrows with bends in them flow smoothly (see Figure 6.6). If you have to move

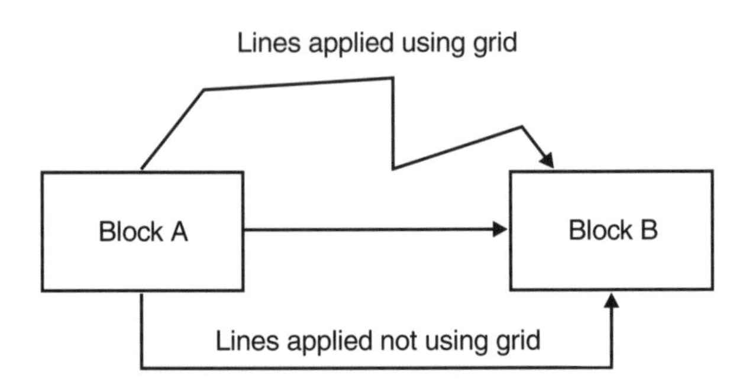

Figure 6.6 Using the grid.

something out of the grid position, first draw it to the grid, then turn the grid off to adjust that specific item, and finally turn the grid back on to capture the new location.

Professionals often draw something very large and then reduce it. This has the effect of reducing imperfections in the final product. By increasing the size of the object as you draw it, you can draw to the grid, making sharp breaks in lines that are supposed to be smooth. Then, when you reduce the object, the sharp breaks will not be noticeable. Try making a complex line drawing on PowerPoint, and you will find this is precisely true.

Draw on the vertical or horizontal and then rotate the object when you are done

By drawing an object with straight vertical or horizontal orientation, you can use the grid to great advantage. You can also easily achieve symmetry. A professional artist can draw something at an angle because of his or her training. Without that training, you will be much better off drawing it at a boring angle and then rotating it with the software.

Tab, don't space

Because most computer-type fonts are proportional, spacing to align items can make them uneven. Use the tab to line up items.

Example of artistic drawing by nonartist

Figures 6.7–6.9 show steps that will allow you to draw a symmetrical and artistically pleasing figure for use in an explanatory graphic slide. This technique can be used to draw just about anything. Remember that if you get it a little bit wrong, you can return to the first step and make minor changes at any time.

Figure 6.7 shows the object the nonartist wants to create and the first framework of the drawing. The framework uses a line of symmetry, boxes, and vertical lines. There are always two boxes (one a copy of the other) used together. One is above the line of symmetry, and the identical box is below the line in the mirror image.

Figure 6.8 shows how the arbitrary line tool is used to connect the parts of the figure. The grid is on, so it is easy to move the cursor to the

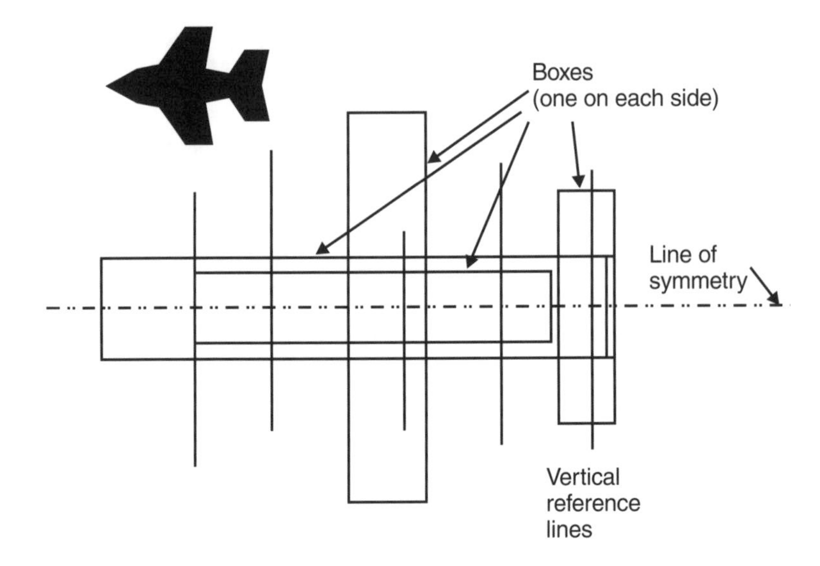

Figure 6.7 Set up the framework for the drawing.

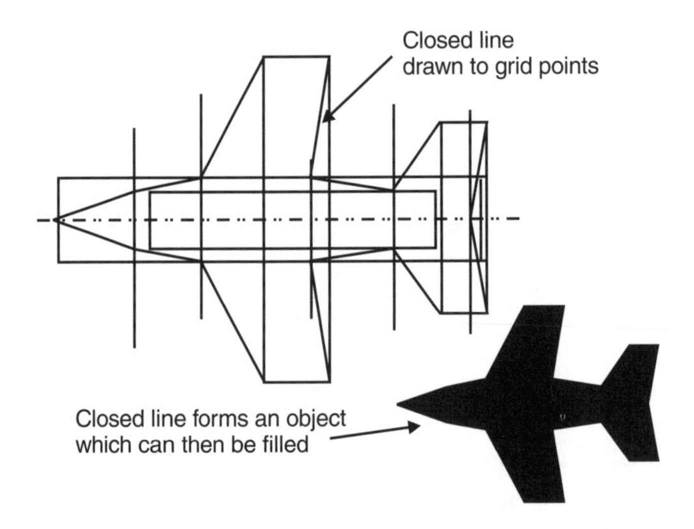

Figure 6.8 Draw a closed line connecting the proper intersections using the tool that allows you to click on a point to continue the line from a previous point.

intersection of two lines and click. This makes a line from the last click point to the current click point. Then, move on the next point. If you make a mistake, just start the line again.

Figure 6.9 shows the line closed in the shape of the desired object. The line will form into an object when any point is double-clicked or when you get back to click on the initial point. This approach can be used to draw almost any simple object for use in effective briefing slides.

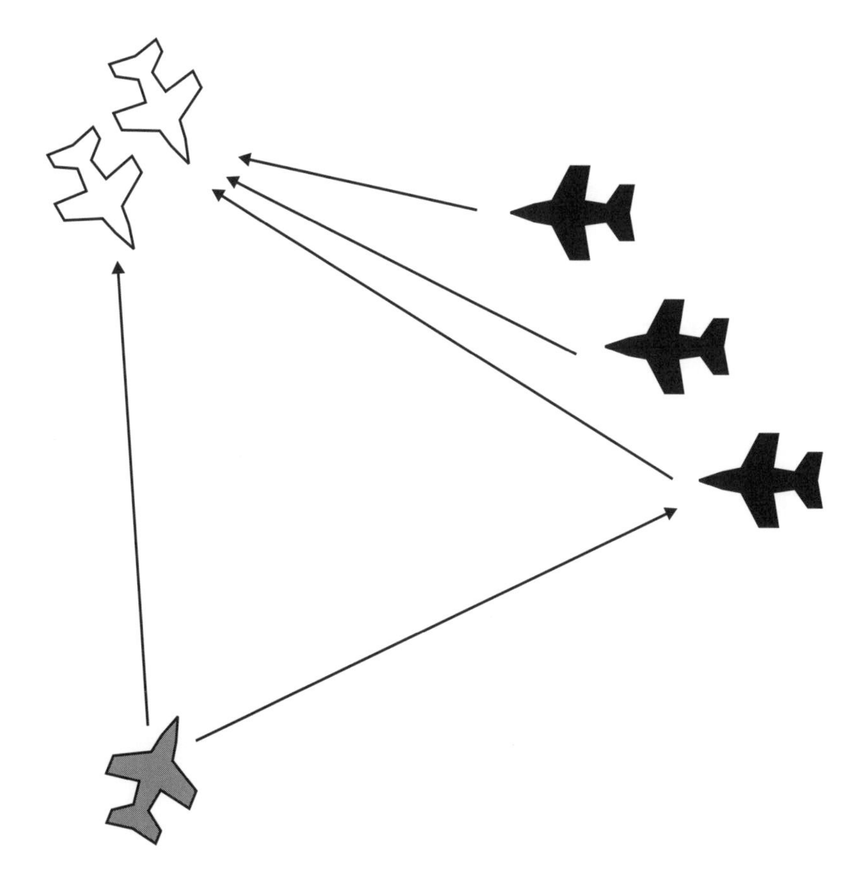

Figure 6.9 Place, rotate, and fill copies of the object to create desired relationship figure.

7

Logistics

THIS CHAPTER DEALS with those mundane details that do not seem important compared with the difficult task of preparing for a briefing and the terrifying prospect of actually presenting it. However, the logistics are extremely important because they directly affect the ability and willingness of the audience to receive the information you want to give them.

Properly handled logistics cannot make an ineffective talk effective, but improperly handled logistics can certainly diminish the effectiveness of what would otherwise have been a well-received talk.

7.1 Briefing room readiness

Whenever possible, take a look at the room in which you are to speak well before the scheduled time for your talk. Notice the room layout, the equipment present, and the location of any important switches. When you arrive for your presentation, this familiarity with the room will make you more self-confident, and that confidence will make your presentation more effective.

There is obviously a second reason to look over the room ahead of time—there might be something wrong. By finding out about the problem while there is still time to fix it with some grace and dignity, you will be doing yourself and your audience a big favor.

Too many speakers arrive at the last minute and spend their audience's time trying to find things or make equipment work or rearrange chairs when they should be starting their presentations. Even though it may not be your responsibility to prepare the room, it is your presentation that will be diminished if the preparation is not done properly. The few minutes you spend in reconnaissance and repair will be well worth it.

When preparation of the briefing room is your responsibility, there are several elements you should consider:

- Scheduling of the room;
- Room layout:
  - Number and arrangement of chairs;
  - Writing surfaces, if required;
  - Electrical outlets;
  - Lectern and speaker's table.
- Equipment:
  - Sound system if the audience will be large;
  - Visual aid equipment.
- Safety and comfort items, including the following:
  - Room lighting;
  - Room ventilation;
  - Routing of wires.

Each of these elements of briefing logistics is covered in the sections that follow.

7.2 Room scheduling

Scheduling the room in which a briefing is to take place is one of those items that seems routine but is not trivial. Simply put, no room, no briefing! As soon as you know that you are to arrange for a briefing room, determine the size of the audience and schedule the appropriate room. Meetings have a pernicious way of occurring at the exact time that the only appropriate room is not available.

Scheduling your meeting room as far as possible in advance will maximize your chances of getting the room you want. It will also enable you to find a suitable alternative or reschedule the meeting if the appropriate room is not available.

In most organizations, meetings are firmly scheduled and entered in a log kept by an administrative person (perhaps the boss's secretary). If your room is secured in this way, your claim on the room is usually quite safe.

7.3 Room layout

The basic requirements for room layout are that all members of the audience are able to see and hear the speaker and see any visual aids used. There is no "best" room layout because there are many different types of briefing situations. However, there are a few basic elements to consider about any room used for a briefing:

- Electrical equipment should be located near electrical outlets if at all possible.
- The lectern should be placed where everyone in the room can see it. If the group is large, the lectern should be elevated on a podium or stage so that the people in the back rows can see over the heads of those in front.
- In general, the lectern should be on the right side of the room (as seen by the audience) because most speakers hold pointers in their right hands. Since the speaker will move from the lectern to a screen or flip chart to point, placing the lectern as described above will avoid the necessity for the speaker to cross in front of the screen to use the pointer comfortably.

- If projection-type visual aids are used, the screen should be placed so that the audience can see it and the lectern at the same time.
- The room should be arranged so that the audience will not be looking at windows behind the speaker.

The suggested room layouts for briefings in Sections 7.3.1–7.3.4 should be considered only as a starting point. They may be adapted to the shape and other requirements of the room.

7.3.1 Layout for a briefing to a small group

As shown in Figure 7.1, a briefing to a small group is normally held in a conference room with a large conference table, around which all of the audience can sit and take notes. The briefer ordinarily stands at one end of the table to make the presentation.

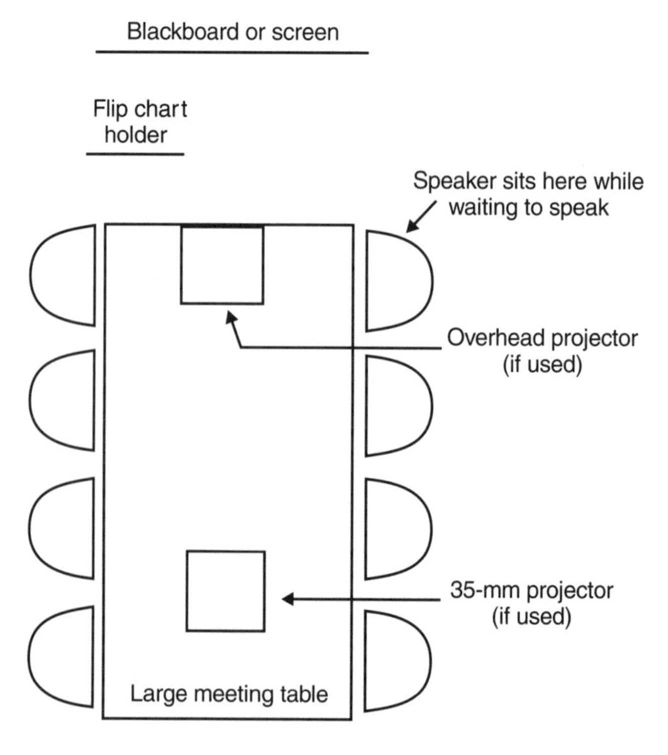

Figure 7.1 Room arrangement for a briefing to a small group.

If overhead projector slides are used as the primary visual aid, the overhead projector is placed on the end of the table, projecting onto a screen at the end of the room. This means that the view of anyone sitting at the far end of the table will be obstructed by the projector, so the chairs should be arranged along the sides. The briefer will change the overhead slides and keep the used and unused slides in piles on either side of the projector.

Thirty-five millimeter projectors require a longer throw than overhead projectors, so if 35-mm slides are used, the projector must be placed farther back on the table. The briefer must either use a remote control to change slides or have a helper seated next to the projector on the left side of the table change them. A movie projector would be placed in the same location.

If a portable direct computer projector is used, it is placed where the 35-mm projector is shown in Figure 7.1. However, many organizations have conference rooms in which the computer projector is mounted on the ceiling. In either case, a remote control unit can generally be used to advance the slides.

If flip charts are used, the flip chart stand should be placed slightly to the audience's left from the center of the speaker's area of the room. This provides the easiest view of both briefer and charts. When both flip charts and a screen are used, the flip chart stand should be placed completely to the left of the screen. Meanwhile, if a blackboard is used, it is ideally placed in the same location shown for the screen.

The chair on the right side of the table, closest to the speaking area, should be reserved for the speaker. This makes it convenient for the speaker to get up to speak and prevents anyone else from sitting there. It is best not to have an audience member in that chair, since the speaker's body may be between that person and the screen when the speaker is using a pointer to make points on the screen.

7.3.2 Layout for a large group at tables

Particularly for long-duration classes and meetings, it is very desirable to place each audience member at a writing surface. In a lecture hall, this is no problem, since every seat will have a fold-down writing surface. However, if you must set up a classroom in a multiple-purpose room, you will have to provide tables for the use of the audience.

Figure 7.2 shows one very effective layout for a large audience seated at tables. If they are available, narrow tables are desirable. They allow the audience to be kept closer to the briefer for easier listening and viewing of visual aids. Most hotels and many restaurants can provide the narrow tables.

An important feature of this layout is that there is a large center aisle. This allows an overhead projector, 35-mm projector, movie projector, or direct computer projector to be used without blocking the view of any audience members. The screen is placed as shown in Figure 7.2. It must be high enough to be seen by all audience members. Ordinarily, the speaker will stand on the right (from the audience's viewpoint), clear of the screen, to use a pointer.

If a lectern is used, it should be placed on the right. It should be placed so that it does not block the screen view of the audience member in the extreme right front-row seat.

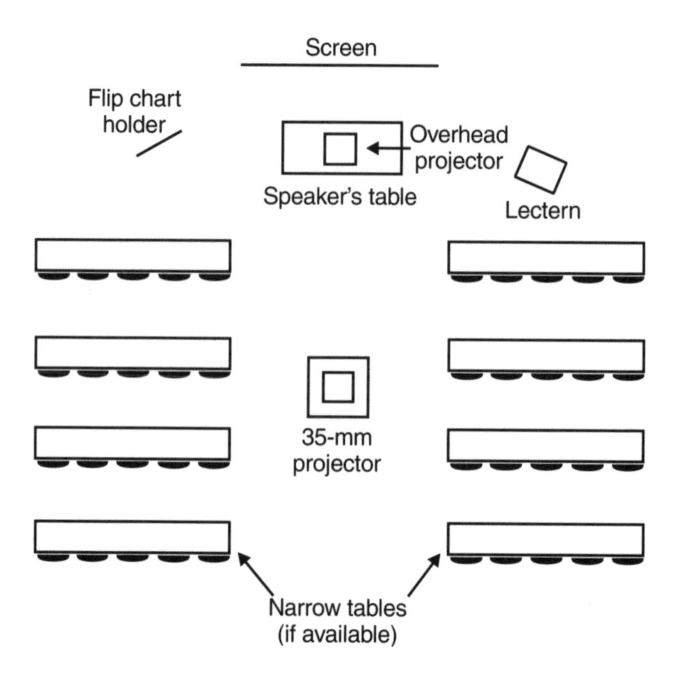

Figure 7.2 Room arrangement for a large group with requirement for writing tables.

If a flip chart stand or freestanding blackboard is used, it should be placed on the left of the room. It should be slightly angled as shown in Figure 7.2 so that it can be seen by all audience members. As a guide, the angle of the screen as seen by the audience members at the extreme front right and left should be about equal.

A small table can be placed in the large center aisle to hold a long-throw projector. If a remote control cannot be used to change slides, a helper can sit behind this table to do that job.

A speaker's table should be provided conveniently close to hold the briefer's material, such as slides and extra handouts. Since this layout is often used for all-day briefings, the speaker may want a pitcher of water on this table.

7.3.3 Layout for large group in theater-style seating

For very large groups (hundreds of people) it is most common to use theater-style seating with rows of chairs. When designing a room layout for this type of seating, the main concern is the logistics of getting such a large number of people safely in and out of the room in the most expeditious manner. For this purpose, it is best to have an aisle down each side of the room and a large aisle in the middle. A large center aisle also provides a convenient area for the location of overhead, 35-mm, movie, or computer projectors without obstructing the screen view of any audience members.

The screen, which must be large, should be in the center of the wall faced by the chairs. If the room is not square, the stage area should be against the narrower wall so that no members of the audience will have to view the screen from too great an angle.

If flip charts are used, they should be placed on the left and angled as shown. It should be pointed out that the use of flip charts with a large audience is quite rare. Because of their size, flip charts would not be readable to the audience in the back of the room.

Ordinarily it is desirable to have a raised stage area when the room is organized for theater-style seating. Otherwise, it is very difficult for audience members in any but the few front rows to see the speaker comfortably. Most large hotels can provide risers to create a one-foot high stage as large as required. This is usually adequate for the audience to see standing speakers, or even a group of speakers who are seated in chairs on the stage.

A permanent stage built into a room is usually three or more feet high. In this arrangement, you may want to eliminate the speaker's table shown in Figure 7.3 and place the overhead projector on the floor of the stage so it will not be between the audience and the screen. With the projector on the floor, the speaker will have difficulty changing slides, so a helper should sit on a chair in front of the stage to change slides for the speaker. Place the lectern at the right of the stage so that it will not interfere with the audience's view of the screen.

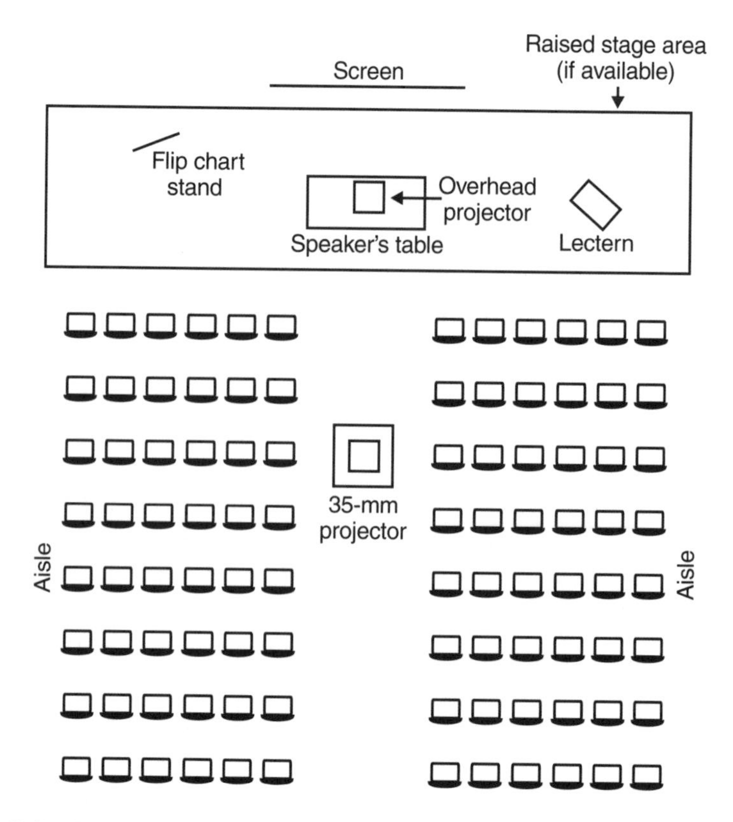

Figure 7.3 Room arrangement for a large group when tables are not required.

7.3.4 Banquet-style seating

The room layout shown in Figure 7.4 is usually the most desirable type of banquet-style seating. It combines the maximum visibility of the speaker with a comfortable dining arrangement.

The layout in Figure 7.4 shows five-foot diameter tables with eight chairs around each. Since round tables normally come in sizes from four feet to six feet, the number of chairs at each table can vary from six to ten. Hotels will usually give you some choice as to the number of chairs to set at each table. If you are not near the room's capacity, you can set the tables

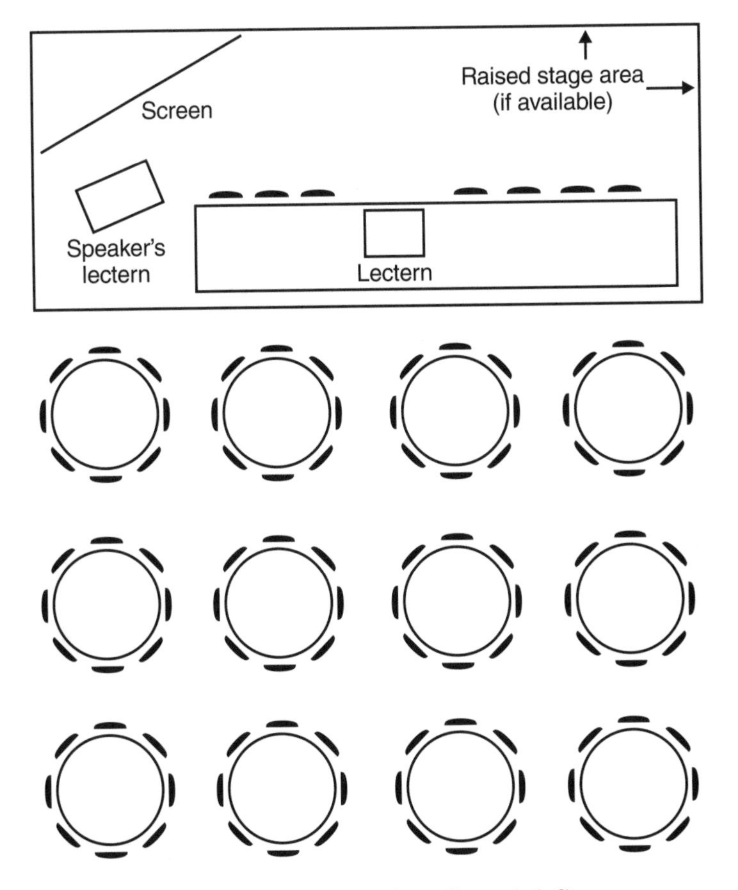

Figure 7.4 Room arrangement for an after-dinner briefing.

with more room for each member of the audience. There are customarily even numbers of chairs at each table.

If a head table is used, it will be along one wall of the room. When the audience is large, it may be desirable to elevate the head table on a stage. There is usually a tabletop lectern placed in the center of the head table for use by the master of ceremonies. Other speakers, who are typically seated at the head table for dinner, may use the same lectern. However, it is good practice to provide a second, freestanding lectern at the left end of the head table (from the audience perspective) for use by the main speaker of the evening. This allows the people at the head table to see the speaker perform.

If the speaker uses flip charts, they are located at the left by the lectern. If any kind of projected visuals are used, the screen would be placed diagonally in the corner of the room as shown in Figure 7.4.

When round tables are used, the people at that part of the table closest to the head table usually turn their chairs around to face the speaker after the meal is over and the presentation starts.

Figure 7.5 shows the classic banquet room layout with a head table and several long tables at which the audience is seated. A lectern at the center of the head table is used by all speakers. This is sometimes required because of custom or because of restrictions on the room facilities.

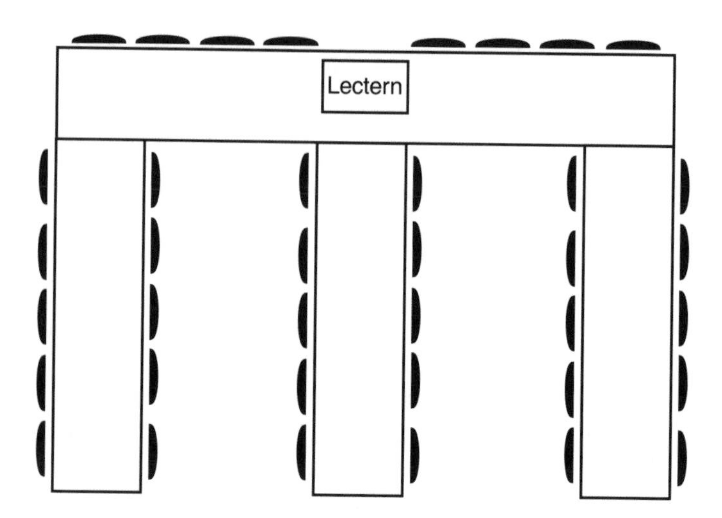

Figure 7.5 Classic banquet-style room layout.

However, it is much less desirable than seating at round tables because it is difficult for the audience to see the speaker. If round tables are not available, small rectangular tables can be placed separately to achieve the same general result. In this type of layout, visual aids are usually awkward, because the audience is facing in three directions. The best compromise is usually to place a screen diagonally in one of the corners of the room behind the head table. Unless the speaker has a remote control, someone else will have to flip slides to allow the speaker to remain at the lectern.

7.4 Collection and preparation of equipment

It is necessary to check with all of the participants in a program early to determine what types of visual aids they plan to use. Once you know what visual aids are to be used, be sure to obtain all the equipment required.

- Projectors of any kind require screens that are large enough for the audience. They may also require electrical extension cords.
- Movie projectors require take-up reels.
- Slide projectors require extra slide trays and remote controllers.
- Flip charts require flip chart stands.
- Blackboards require chalk (colored chalk is a nice added touch) and erasers.
- Whiteboards require the proper pens and erasers.
- Most types of visual aids require the speaker to use some sort of pointer. If the screen is large, a light pointer (which projects either a laser beam or a bright white arrow onto the screen) is required.

Even when you have gathered the equipment, the job is not done. You need to be sure it all works—far enough in advance to fix or replace anything that does not. Consider the following:

- Do the projector lights work? Have extra bulbs available.
- Does the slide projector properly advance slides? Does the remote controller operate properly?

- If a direct computer projector is used, is it compatible with the medium holding the briefing (floppy disk, high-density disk, CD, etc.)? Does the computer driving the projector have the proper software? Does the remote controller work properly?
- Does the screen set up properly? Be sure that there is not a strong wind from an air-conditioning duct rippling the screen.
- Is the flip chart stand sturdy enough to avoid falling over when the charts are flipped?
- Do the felt-tipped pens work? The pens used on whiteboards dry up very quickly when the caps are left off.
- Is there really chalk in the box of chalk by the blackboard?
- Is the blackboard or whiteboard clean? Is the eraser clean?

7.5 Safety and comfort factors

Any time that a large number of people are assembled in a single room, there are several safety and comfort considerations that have overriding importance.

- Be sure that fire exits are clearly marked and are not obstructed by furniture or equipment.
- The room layout must provide for aisles that are adequate in number and wide enough so that the audience could be quickly evacuated in case of emergency. This will also minimize the time it takes to get the group into their seats at the beginning of the program and out at the end.
- Be sure that there is adequate ventilation in the room. Windows should be opened or the air conditioning turned up to handle the increased load of many people in a room.
- Do not forget about restrooms and drinking fountains.
- If possible, make a firm policy about smoking in the meeting room—no smoking, smoking "OK," or smoking only in some sections of the room. Many organizations now have policies against

smoking in any conference rooms. If not, you will be wise to organize the room into smoking and nonsmoking sections and to mark them clearly.

- Tape down all cords to prevent tripping accidents. This includes extension cords, projector remote control cords, and long microphone cords.
- Use proper extension cords that are in perfect condition. It is a good idea to tape the outlets of extension cords onto the plugs of equipment to prevent them from separating and causing a shock hazard.

7.6 Marking of visual aids

Before (not after) they are dropped, be sure that all visual aids are properly marked. All slides must be numbered in a consistent manner so that they can be quickly placed in order and properly oriented if they become disorganized for any reason.

One excellent technique that is often used for 35-mm slides is to place the slides in a stack in the proper sequence and then draw a diagonal line across one side of the stack of slides as shown in Figure 7.6. This can also be done with overhead projector slides (if they are mounted in frames) and with stacks of display cards.

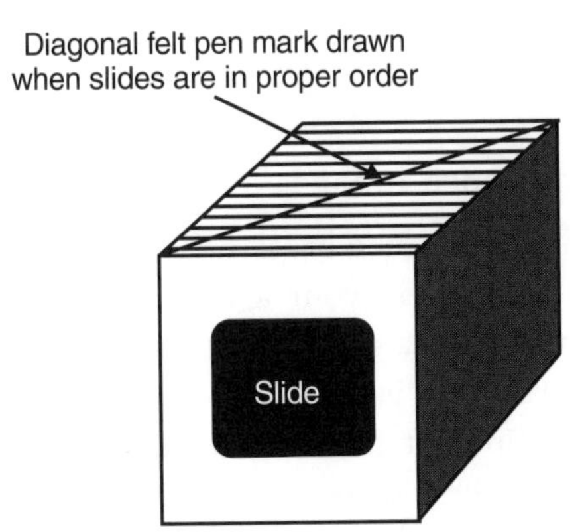

Figure 7.6 Set of slides marked for easy arrangement.

In addition, carefully mark the disks containing briefing material. When modifications have been made, be sure that the disk contains the latest version. In general, it is best to delete all older versions from the disk to prevent confusion.

7.7 Multiple-speaker logistics

If you are responsible for a meeting in which several people will be giving briefings, there are a few special logistics that must be considered. These involve getting people on and off the platform on time, getting their visual aids from them and back to them, and handling questions and answers. These logistical considerations will be covered in detail along with the larger subject of managing the multiple-speaker briefing in Chapter 9.

7.8 Feeding people

After all is said and done, people are still people. When they get hungry or thirsty, they lose interest in intellectual activity, such as technical briefings.

If you are responsible for a meeting of three hours' or more duration, you need to consider coffee breaks or meals or both. If you are responsible for a mealtime meeting, the feeding of the audience must, of course, be part of your planning. The information in Sections 7.8.1–7.8.3 will be helpful to you in either case.

7.8.1 Coffee and lunch breaks

Ideally, you should plan a coffee break halfway through each half day of your meeting and show it on your agenda. However, because coffee breaks and lunch breaks take a significant amount of time (see Table 7.1), they cause major disruptions in the flow of the subject matter. The minimum suggested times in the table allow the extra time it takes to move a group into and out of their seats. If you schedule breaks shorter than suggested, you will usually find that you must delay the restart of the

Table 7.1

Schedule Time Required for Coffee and Lunch Breaks

Size of Audience	Minimum Coffee Break Duration	Minimum Lunch Break Duration
Less than 12	10 minutes	30 minutes
12–50	20 minutes	45 minutes
Greater than 50	30 minutes	1 hour

meeting while people are getting settled into their seats. To preserve the coherence of the meeting, each quarter-day period between major breaks should contain some logical subdivision of the material presented. This is true for half-day or full-day briefings.

For coffee breaks, you need to consider how your audience will get from their chairs to the coffee and rolls and how they will be served. If people need to purchase beverages and snacks from vending machines, you need to allow extra time for this activity. If coffee and rolls are provided, you will need to make provision for disposing of used cups, napkins, and other waste.

For lunch breaks, you need to consider carefully how people are to get their food. You can shorten the lunch break somewhat if you are planning to hand out box lunches. For a served, sit-down lunch, the times listed in Table 7.1 assume a fairly efficient meal service and a short walk from the meeting room to the meal room. Add extra time if either of these conditions does not exist.

If the weather is warm, many meeting planners like to provide cold soft drinks rather than coffee for afternoon "coffee breaks." There should always be restrooms and, if possible, telephones in the area where your audience will be taking its coffee breaks.

You can expect people to return from coffee breaks refreshed and ready to absorb more technical material. However, lunch is a different matter. For the first hour after lunch, people have a tendency to be drowsy. If your day of briefings needs to include any "walk around" activities—such as tours of facilities or demonstrations of equipment—right after lunch is an excellent time to schedule them. Movies are also very nice right after lunch if they are not too critical to the day's subject matter. Those who are going to nod off will do so while the room is dark and will wake up after the film refreshed and ready for the rest of the day.

7.8.2 Breakfast or lunch meetings

A great deal of business is conducted over breakfast or lunch. Although formal briefings are sometimes held while the audience is eating, this normally does not work too well. It is much more common (and successful) to break from briefings and conduct informal discussions while food is actually being consumed.

Because the activity of eating requires a special concentration—to select, prepare, serve, and keep from spilling food—it is very difficult to give proper attention to a technical briefer at the same time. For audience comfort and briefing effectiveness, it is suggested that mealtime briefings be started after the audience members have finished their food. If time is pressing, you can usually start while they are finishing their coffee.

7.8.3 Dinner meetings

Any meeting at which a briefing is to be given along with a major meal is just like any other mealtime meeting but more so. The meal is more formal and takes more time, and people expect to have pleasant, mealtime conversations with those around them. One formula that is often used is to make any necessary introductions before dinner and start the formal briefing after the dinner dishes have been cleared and coffee has been served.

7.9 Handout material

There are two logistical concerns about handout material: First, be sure that you have enough copies, and, second, pass them out with a minimum of confusion.

People are sometimes offended when they do not receive a personal copy of the handout material, so it is a good idea to have a copy for everyone you know will be present and have reserve copies (about 10% more). Also, it is a good practice to mark a special copy for yourself. You can make notes on your copy to help you answer questions about the origin of the data presented and other peripheral but important matters.

In most speaking situations, you should hand out your written material after your talk so the audience will be listening to you speak rather than reading the material during your briefing. If it is a small group, you can simply pass around a stack of copies after your talk. If the group is large, you can minimize the disruption to the meeting by handing premeasured stacks of copies to people at the end of each row in the audience—with instructions to "take one and pass them down." A better approach, if it is permissible, is to have the audience members pick up their own copies as they leave the room for the next coffee break.

If you want the audience to have the material in its possession during your talk, pass the copies out before your briefing. Then give the audience a few moments to rustle the pages before you start talking. The alternatives are not good: Passing out copies during your talk will be very disruptive; if you ask people to pick up copies on their way into the room, a few will forget to do so and will probably interrupt you to ask for copies.

7.10 Ten "little" items that everyone forgets

It is the little things that always seem to get us in trouble, so here is a list of 10 seemingly insignificant items to think about. Although it is obvious that all of them do not apply to every type of briefing, everyone forgets them sooner or later.

1. Pointer;
2. Chalk;
3. Extension cord;
4. Extra bulb for projector;
5. Special pens to write on slides;

6. Plug adapters for demonstrations;
7. Sufficient copies of handouts;
8. 35-mm slide trays compatible with projector;
9. Projector remote controller;
10. Projection screen adequate for audience.

8

Presentation Technique

THE OTHER CHAPTERS of this book deal with the details of planning and preparing for briefings. These skills can be learned from books. However, no matter how well you plan and prepare a briefing, it is not a briefing until you stand up and deliver it to an audience. That is the scary part, and unfortunately, it is *not* learned from books.

Just as riding a bicycle cannot be taught, effective presentation technique cannot really be taught. You must learn by doing it. This bicycle-riding analogy is very appropriate. Any skilled bicycle rider can *show you how* to ride but probably cannot *tell you how*. Even if someone did explain to you the tiny shifts in weight that you must make to balance a two-wheeled vehicle, it would not help you ride the bike. You just have to climb aboard and try it. Eventually, after a few skinned knees, you become a skilled rider—and probably will not remember how you learned. You did learn, and, just as certainly, you will learn to be an effective public speaker if you persevere while observing some rules of good practice.

To start, consider what constitutes excellent presentation technique. This is best done by dissecting the performance of expert briefers you have observed.

- Excellent presenters seem to speak effortlessly.
- They do not seem to be nervous.
- They make frequent eye contact with the audience.
- They exude enthusiasm about the subjects they present.
- They seem like nice people.
- They seem to know what they are talking about.
- They can be easily heard, and the audience follows their visual aids without any difficulty.

No one is born with these traits, nor can these traits really be directly taught. However, you can learn them through observation of good speakers and by emulation of their technique, while you give as many briefings as you can.

Only practice will make you a polished public speaker, but there *are* a few tricks of the trade that will help you look much better from the start and will help you become polished much more quickly. The purpose of this chapter is to make you aware of those tricks of the trade; the practice that will polish your skills is up to you.

After you study the tricks, go back and review the above traits of a good speaker. You will see that good speakers have internalized the tricks and now use them naturally and fluently, achieving the traits you have observed. Like table manners, the mannerisms that make a public speaker excellent are not impressive until they have been used enough to become automatic. The tricks fall into four categories: generalities, things to do ahead of time, habits to avoid or break, and ways to speed your improvement.

8.1 Generalities

Like most generalities, these hints about presentation technique will mean different things to different people. You should implement them in the way that is best for you. Also, it may take you a while to learn to do *all* of these things. Still, they will make you a better presenter if you use them, and you should always do the best you can.

8.1.1 Speak in a conversational tone

If you speak to your listeners in a normal tone of voice, they will be more comfortable with you and with the material you bring to them. Naturally, if the audience is large, you will need to use a microphone to make yourself heard without shouting.

Please do not confuse conversational tone with monotone. In everyday conversation, you use a great deal of variation in pitch, volume, and intensity. However, you use these qualities in normal context and without straining, as you do when shouting or whispering.

This tone of voice makes the audience feel that you are relaxed and confident, whether you are or not. Your increased confidence will make the audience more receptive to the ideas you are presenting.

In earlier days, when there were no sound systems to help, outstanding speakers used oratorical style to project their voices. This sounds very impressive, but it cannot compare with a normal conversational tone for the effective transfer of information from a speaker to an audience.

8.1.2 Stand beside your visuals

No matter how good your visual aids are, the audience will not benefit from them if they cannot see them. It is true, "You make a better door than a window."

Stand at the side of your visuals when using a pointer, so you do not block the audience's view. If you are right-handed, stand so that your visual aids are to your right. This way, you can point to parts of your visual aids without reaching across your body. It will be natural for you to face the audience while speaking, and you will have the minimum chance of blocking any audience member's view. Figures 8.1 and 8.2 show right and wrong ways to use your pointer.

8.1.3 Use a pointer

Use some kind of pointer when presenting visual aids; do not point with your finger. When pointing out items on a projection screen, a pointer

Figure 8.1 Proper position for use of pointer.

will allow you to reach all parts of the screen without having part of the picture projected onto your face and body. A pointer also allows you to keep a normal, balanced stance while reaching the top, far corner of the screen. This makes you look more confident and professional.

When pointing out items on an overhead projector, use a pencil or similar object for a pointer. A pencil has a nice sharp point and can be left in place on the slide until you are ready to move it to the next point. If you use your finger, it will project onto the screen along with the slide material. A projected finger silhouette is huge and blunt. Besides looking amateurish, the silhouette makes it difficult for the audience to see exactly where you are pointing on the slide.

Move the pointer with slow, deliberate motions. Remember that the pointer tells the audience where to point their eyes, to see exactly what you are talking about at the moment. If you move the pointer too quickly, the audience's eyes will take time to catch up. This is both distracting and uncomfortable.

Figure 8.2 Trying to use the pointer from the wrong side.

Laser pointers are very appropriate for large briefings because of the large size of the screen. There are now some very small and inexpensive laser pointers on the market. They are bright enough for the audience to see, but their small size makes it difficult to hold them steady. If you hold a laser pointer with a shaking hand, the whole audience will know (or think they know) that you are nervous. Thus, it is a good idea to use a small laser pointer from a braced position. Rest your hand or arm on the lectern. Another technique is to use the laser pointer in bursts. Hold it on your point for a few seconds, then turn it off until you make the next point.

8.1.4 Speak to individuals in the audience

An audience, taken as a group, is a sea of faces. Individually, they are *people* who are reacting to your talk. By speaking to a few individuals in the audience—focusing on one person for a few seconds and then moving to another in another part of the audience—you will be able to achieve the same conversational attitude you have in a casual conversation with two or three people.

A caution: do not move your eyes constantly across the audience. This is distracting, both to you and to the audience. The pauses to talk to one individual for a few seconds make the difference.

8.1.5 "Read" the audience

While you are speaking to individuals in your audience, you can observe what they are telling you by their body language. As you know from years of one-on-one conversations, an individual's eyes sort of glaze over when he or she has stopped listening to what you are saying. The facial expression will also tell you when someone does not understand something you have said or strongly agrees or disagrees with a statement. All of this allows you to communicate more effectively by listening with your eyes while you are speaking. This reading of faces becomes particularly valuable in public speaking because audience members cannot (politely) convey their moment-to-moment reaction to what you are saying in any other way.

If you want to have even more tools with which to read the audience, study the ways that people unconsciously communicate with other parts of their bodies. Nierenberg and Calero's excellent book *How to Read a Person Like a Book* describes a number of specific body positions that consistently indicate how one individual is really reacting to another individual—even if the words being spoken do not tell the same story.

Individuals in an audience give the same facial-expression and body-position clues to their feelings as they do in one-to-one interactions. By reading these indicators from several individuals spread through your audience, you can gain such valuable information as the following:

- The audience is confused by an explanation you are presenting.
- The audience cannot hear you.
- You are speaking too loudly.
- The audience is bored and needs a change of pace.
- The audience agrees (or disagrees) with what you are saying.

8.1.6 Present your visuals with flair

When you expose a new visual frame to your audience, either by changing slides or turning a flip chart page, start talking about the visual frame a few seconds before the audience can see it. For example, your introductory sentence for a visual frame might be, "As you can see from a top view of the aircraft, the antennas are flush mounted." This technique would have you expose the visual showing the top view about halfway through the sentence.

This technique will give a quality of smoothness to your talk by easing the audience across the discontinuities in visual information that occur when you change the visuals. If you stop talking, then plunk down a new overhead slide, and then start talking again—your audience will be halfway through reading the slide before you start. Audience members will have to either ignore your talking and finish reading the slide or stop in the middle of their reading to return to the beginning of the material where you want their attention. The "start talking, then change visual" technique keeps the audience with you right from the start of the visual frame.

While a visual frame is exposed to the audience, do not read it—talk about it. The audience members will read the visual themselves; your job is to add information to the framework of the visual while they are reading it.

8.1.7 Answer questions effectively

Audiences are impressed with speakers who can handle questions well. This, like all other briefing skills, can be learned. The way to start is to use the following procedure:

- Listen closely to the question. Be sure you understand the question before you start to answer.
- Repeat the question if everyone in the audience cannot hear the questioner's voice. Otherwise, the audience will hear a disjointed answer from you and will not be ready to benefit from the knowledge you pass along in the answer.
- If the question is poorly stated or too complex, rephrase it for clarity before answering. This will let the audience know exactly what question you are answering.

- Answer questions as directly as possible. It is all right occasionally to go off into a detailed explanation of a subject that is shown to be important by the content of a question. However, if this is done very often, it quickly uses up the question-and-answer period and deprives other audience members of the chance to ask questions that are important to them.
- If you do not know the answer, do not bluff. Admit that you do not know and offer to get the answer back to the questioner at a later time. If you try to bluff your way through with an answer that sounds good, you will almost always get caught. Someone in the audience will know the real answer, and if you are wrong, that person will probably give you the full benefit of that knowledge (in a loud voice) while your wrong answer is still hanging in the air. There is no shame in not knowing everything. No one does.
- Use visual aids to help answer complex questions. If you have used a visual aid that contains the seeds of the answer, bring the visual back up and build on it. An experienced briefer will prepare backup slides in important areas that are slightly beyond the scope of the briefing but may well come up in questions.

8.1.8 The big secret

Unlike most of the other subjects covered in this book, there is a big secret to effective public speaking in technical briefings. Using the big secret will make you a public speaker to whom audiences like to listen. That secret is *enthusiasm.* Make the effort to convince yourself that the message you are presenting is important to you and to the audience. Then let your enthusiasm show on your face. Even during your very first briefing, when you may be nervous and uncertain of yourself, the audience will sense your excitement with the subject matter and will respond favorably to you.

Conversely, if you do not care about the subject, why should the audience be interested in what you have to say? Be excited! Show you care.

8.2 Things to do ahead of time

It is always wise to prepare for future times of need. When you are making your presentation, you will be very busy and probably nervous. It is a good idea not to trust yourself with any more detail than is necessary. Do whatever you can ahead of time, leaving to be done only what you *must* for the actual time of need. In Sections 8.2.1–8.2.5 you will read about specific things you can do ahead of time to make it as easy as possible for you to present your material in an excellent manner.

8.2.1 Dress the part

"People listen with their eyes," said my high-school band director. The first thing that the members of your audience will experience from you—before you even open your mouth—is the way you dress.

Some briefing situations require specific items of clothing, such as uniforms or safety clothing. However, in almost every professional briefing, the "authority look," as John Molloy described in his book *Dress for Success* will make the best impression on your audience. Molloy gives examples of proper authority clothing for men and women. He presents research results that support the selection of specific combinations of clothing, colors, and fabrics that make people believe that you know what you are talking about until you open your mouth (to convince or not). Wearing authority clothing when you speak will not only make your presentation more effective, but will also leave the audience members with a more favorable impression of you as a competent professional.

Any dress shoes will work if you are lecturing for an hour, but the all-day lecturer needs to take special care with footwear. Wear shoes that provide some degree of cushioning and excellent arch support. After eight hours of lecturing in typical dress shoes, you'll feel your heels coming up through your knees.

8.2.2 Be familiar with your visuals

Take the time to become very familiar with your visuals before your presentation. Be comfortable with the material on each visual frame and know the order of the visuals so well that you instinctively know what is on the next few frames. This will come from rehearsal with your visuals, but it can also be gained by flipping through your visuals several times.

If you are using visual aids prepared with a presentation software package, you will probably be able to make a set of reduced visual aids—six or more to the page. Although the individual slides will be too small to read well, you will be able to see the order at a glance. Therefore, you might want to have a set of reduced visuals available to you during the presentation.

8.2.3 Rehearse your talk

No matter how well you know the subject, your presentation will be smoother and you will be more comfortable with it if you rehearse. During rehearsal, you will work out the way you want to say things and get a feeling for the timing of your talk. You will also discover tiny changes you want to make to improve the talk.

If you practice using a speaking outline, your eyes will return naturally to the proper point on the page so that you will be more comfortable with looking up at the audience. This will make it far easier for you to make good eye contact.

8.2.4 Know thyself

By the time you have given one or two talks, you will discover that you have a few traits that negatively effect your presentation and can be controlled by proper action before you start. The following are a few examples:

- The key jingler: Some speakers habitually place a hand in a pocket to jingle keys or coins while they are talking. This is distracting to the audience. If you are a key jingler, you can easily avoid the problem by emptying your pockets before you speak. Also, because you will find no keys to jingle when you try to do so during your talks, you will probably stop putting a hand in your pocket to search for them.
- The dry-throat cougher: If your throat gets dry while you are speaking, do not feel lonesome. Everyone who speaks very long gets a dry throat sooner or later, and most inexperienced speakers have that problem until they get nervousness under control. This is easily fixed by taking a small drink of water or sucking on a cough drop shortly before you get up to speak.
- The pointer torturer: One very popular kind of pointer collapses back into itself so you can put it in a pocket like a pen. Some speakers spend their whole time at the lectern extending and collapsing their pointers. The audience soon becomes fascinated with the

pointer exercise and stops listening seriously to anything said. If you are a pointer torturer, avoid collapsible pointers—use solid wooden ones.

8.2.5 Keep your throat working

An inexperienced speaker can develop a dry throat from nervousness (or perhaps hyperventilation) even during a very short presentation. Even the most experienced speaker's throat will dry out in a one-hour talk, and teaching an all-day course raises the problem to the level of a military campaign. The answer, of course, is to drink water.

Ice water is often provided to speakers and works fine for a very short talk. However, to keep your throat open in a longer talk, drink room-temperature water. A small bottle of water at the lectern is less likely to be spilled than is a glass of water. In general, a "sport top" on the bottle causes you to work too hard for your drink, causing you to sip daintily. Don't be dainty—slug that water down. You want to take a large quantity of water to replace the water you are exhaling and to keep your mucous membranes moist in the face of all that hot air you are flowing over them.

If you are lecturing all day, take some extra measures: Start your serious water drinking a day or two ahead of the lecture. Keep track of the amount of water you are drinking; two or more liters of water a day in addition to the fluids you drink at meals and breaks is not at all out of line if you are lecturing. Another measure used by professional "marathon talkers" is to drink tea with honey or glycerin (available in drug stores in Europe) during breaks. Some professionals eat a high-fat breakfast before a lecture day (presumably to lubricate the throat).

Be careful of cough drops. Most contain menthol, which will irritate your throat if used over a prolonged period of lecturing. If you find some that do not have menthol, they may work for you, but good old water is the preferred solution for most professional speakers.

8.3 Things to avoid

This is an incomplete and completely unscientific list of bad habits to break or to avoid acquiring if you want to be an effective public speaker.

They are common among technical briefers, and they drive audiences crazy. They interfere with the communication process by annoying the audience and drawing its focus away from the material being presented.

- Repeated phrases: Common examples of the overused phrase are "OK" and "You have your. . . ." Some speakers use their favorite phrases at the beginning of every sentence. By the end of such a talk, the audience wants to scream.
- Many rhetorical questions: Some speakers will start every outline point with a rhetorical question, then answer it. For example, "Now, how would you start up the system? . . . The sequence of actions is turn on the power, load in the boot-strap tape, . . . "
- Physical threats: Listeners are in no mood to accept technical information if they feel physically threatened by something the speaker is doing. Here are several examples from actual briefings in which a physical threat stopped the audience from accepting any information because the listeners feared for their own safety.
  - A speaker was supposed to be teaching safety in handling firearms. He demonstrated with a large (presumably unloaded) pistol that he occasionally pointed at the audience.
  - A speaker was using a laser pointer to indicate points on slides on a very large screen and occasionally waved it near the audience.
  - A talk about missile guidance involved demonstration of a gyroscope that was spinning very fast. The demonstration gyroscope made an ominous noise like it was about to fly apart at any moment.
- Fiddling: Any repeated physical action can become distracting, including pushing glasses up on your nose, buttoning and unbuttoning your coat, playing with the microphone, or thumping the lectern.
- Turning your back on the audience: Some speakers spend much time facing away from the audience. The extreme is the professor who writes on the board with one hand, erases with the other, and mumbles into the blackboard. Other examples are speakers who

face the screen while using their pointers. You should face the audience as much as possible while speaking.

- Speaking in bursts: There is a style of speaking popular among military briefers in which short blocks of information are transferred, and the speaker pauses after each block. The blocks tend to be of constant duration, perhaps one or two sentences, and delivered in a fast monotone. Computers like to get their information that way, but people (even soldiers) like more variety in format.
- Speaking too quickly: One speaker was described accurately as speaking 200 words per minute with gusts to 280. Many new speakers speak more quickly because of nervousness. Slow down to your normal conversational speaking rate so the audience won't have to strain to keep up.
- Speaking in a singsong manner: You have heard speakers who start each sentence at a high vocal pitch and change to a low pitch at the end of the sentence. Then they pause and start the next sentence at the same high pitch. Do not do that.

8.4 Ways to speed your improvement

While it is true that you can become a polished speaker only by speaking, there are some things you can do to speed up the polishing process.

8.4.1 Notice what other speakers do

Every time you attend a technical briefing, notice and make notes to yourself on what the speaker did well and what the speaker did poorly. Then, consciously try to emulate the style of good speakers and to avoid the specific weaknesses of poor speakers.

8.4.2 Record your talks

By taking an audiotape recorder along every time you give a briefing, you can make a private record of your speaking rate, word usage, and vocal variety, among other characteristics. Later you can sit down and review the tape to evaluate your own talk from the point of view of the audience.

Video recording involves much more logistical difficulty than audio recording, so it is not practical to video record every talk. However, you should never pass up an opportunity to have a talk video recorded because it allows you the unique opportunity to experience your talk as your audience saw it. You will be able to see if you are doing any of the "things to avoid," and you will be able to review your facial expression, handling of visual aids, and gestures.

8.4.3 Speak as often as possible

In general, frequent practice is more effective than infrequent practice in learning any physical skill like a public speaking delivery technique. When a long period of time elapses between practice sessions, it is difficult to remember what you did wrong the last time and were going to correct this time.

Any kind of public speaking will help you develop better delivery technique; it does not have to be technical briefings. Many public speaking opportunities arise in service clubs, religious organizations, and other volunteer activities.

If you have much difficulty with delivery technique or simply want to improve as a public speaker, you should consider joining a public speaking club such as Toastmasters International. In a Toastmasters club, approximately 30 members meet weekly. At each meeting, each member gives a short, impromptu speech in response to a question just heard. Then, a few of the members (three or four) give formal speeches that are publicly critiqued by other members. Each member gives a formal speech every month or so, using manuals of speaking projects developed by the parent organization.

Toastmasters clubs are available in every large city in the English-speaking world, and in most small towns. Also, many companies have in-house Toastmasters clubs that meet in the cafeteria during lunch periods.

9

Cruel and Unusual Circumstances

NOT EVERY BRIEFING takes place in a well-equipped conference room. Not every audience is ideally suited to the speaker's abilities or assigned subject matter. Sometimes things go wrong.

Occasionally, for the best of reasons, you must speak in challenging circumstances. This chapter deals with the extra techniques and insights you will need to make your presentation retain its effectiveness in those challenging circumstances.

We will cover speaking in a restaurant setting, working with a translator, the nontranslated talk to a nonnative-language speaking audience, dealing with things that go wrong during a presentation, and dealing with hostility of any kind from someone in the audience.

9.1 Speaking in a restaurant

Restaurants are ideal places to eat food; they are not ideal places for public speaking. A restaurant talk will normally occur in a special meeting room. However, sometimes you must speak to a group at a large table over in the corner of the restaurant itself. In any case, you will be speaking in a situation that has more distractions than you would expect in a conference room. The meeting rooms in hotels are very similar to the meeting rooms in restaurants, except that they can be configured more like normal conference rooms if the meeting format calls for it.

9.1.1 The audience

The most common situation is a talk after a meal. This means that you can expect waiters and waitresses to be clearing dishes, pouring coffee, or both. The audience will have just finished a meal and will still be sipping coffee or perhaps finishing dessert when you start speaking.

9.1.2 Room setup

Normally there will be a lectern from which you can speak. Sometimes, it is freestanding, in the corner of the room, but most often it is a tabletop lectern in the middle of the head table, so that the audience will see you in the middle of a line of people. The head-table audience will have a side view of you. The lighting in a restaurant, even when in a special meeting room, is often not very good, so the audience may have some trouble seeing you.

9.1.3 Visual aids

The limitations on restaurant lighting have the greatest impact on the kinds of visual aids that can be used. Since few restaurants provide visual aid facilities, you must rely on bringing your own aids to the meeting. Either bring them with you or arrange with the meeting host to have them there for you.

If the room light is fairly bright, and there is no way to dim the lights, you are fairly well restricted to the use of flip charts or large stiff cards on an easel. This is always the case if you are not in a completely separate room from the main restaurant.

In a separate room, the lights can usually be turned off, even if they cannot be dimmed, so 35-mm slides or direct computer projection with a remote controller are good choices. After a meal, the use of 35-mm slides with fairly high slide change rates is recommended because they support a faster paced presentation than most other media.

Another advantage of 35-mm slides or computer projection is logistical; they are usually much more convenient to use in a restaurant setting than are other types of visual aids. By using a long extension on the remote control cord, you can change your own slides from wherever you are in the room, even from behind a lectern placed at the center of the head table. If the screen is placed in the corner of the room, everyone (including the head table) should be able to see the slides, and everyone (except the head table) will also be able to see you while you speak. (Proper room setup is discussed in detail in Chapter 7.)

The effective use of overhead projector slides requires some conditions that are often hard to achieve in a restaurant setting. To be able to change your own slides, you need to be physically located at the projector with the screen behind you while speaking. This means that you must have a relatively large area from which to speak, which is often difficult to achieve in a restaurant. If you must for some reason speak from the center of the head table, this arrangement almost never works.

9.1.4 Approach to subject matter

The mood of the members of the audience after they have just had a meal deserves special consideration. When you speak, they will have just eaten, and their bodies will have the digestion of food as a priority. However, this also varies with the time of day and the nature of the meal. If the meal has been a light or moderate breakfast, you can give any kind of a technical talk you like—the meal will have little impact because people are used to working right after breakfast.

A fairly light lunch (one sandwich, no alcohol) is close to the light breakfast in the way it affects your audience, but a heavy lunch is almost like the evening meal in its effect. After an evening meal, members of your audience have had a personality change—their intellectual awareness has been dulled. They are digesting food, and in many situations have had at least a little alcohol. In this situation, they must be approached on an emotional basis. That does not mean that you should not present a technical subject, but it means you must approach it in an emotionally oriented way.

To start, be aware that the audience will have a tendency to nod off (literally) if things drag. The comments in Chapter 8 about reading the audience's eyes are very important here. Watch three or four people. When their eyes start to glaze over, it is a warning that they are about to fall over. Therefore, you need to pick up the pace of your briefing. You have spent enough time on that visual aid or the particular point you were making. They have stopped receiving, and you should stop sending.

In every talk you should show that the subject is important to you. In a restaurant situation, however, you need to make that importance to you or your audience much more the focus of your remarks—with less intellectual-level backup material. The audience in the restaurant situation will be both more accepting of your conclusions and less patient with the "work" of taking in the technical details that prove them.

9.2 Speaking through a translator

When giving a talk to an international audience, you may have to use a translator. There are two types of translation, simultaneous translation and delayed translation. Both present unique challenges to the speaker.

9.2.1 Delayed translation

For nonsimultaneous translation, someone who speaks both languages will listen to a short passage of what you have to say and repeat it to the audience while you wait. This takes patience and understanding on both the part of the audience and the speaker to make it work.

First, as the speaker, you will need a good set of notes from which to speak so you will not forget where you are in the talk. Second, you need to pause for translation often enough so that the person translating can remember what you have said. The best approach is to complete each thought in a single sentence and then pause. Give the translator enough material to make a complete statement, but only on a specific subject. Pausing in the middle of a sentence gives the translator a bad problem because the sentence structure in different languages may vary greatly. For example, in English, the verb is often near the beginning of a sentence, while in German, the verb is typically at the end of the sentence. If you pause halfway through the sentence, the translator may have to wait for you to finish before the German can be constructed.

Even though the translator will be delivering the translated speech, speak directly to the audience. In most situations, some of the audience will have enough knowledge of your language to understand part or all of what you are saying. More important, the audience will receive the nonverbal part of your message (gestures, voice inflection, and particularly facial expression) directly from you. No translation is needed.

While the translator is speaking

Pretend that you know what the translator is saying, even though you cannot understand a single word. Face the translator and listen intently.

9.2.2 Simultaneous translation

A much more comfortable situation for both the audience and the speaker is provided by simultaneous translation. Simultaneous translators are real experts who listen to your talk and speak to the audience in their own language with a delay of only a second or two.

The translator sits in a soundproof booth, and the audience listens through earphones. You speak to the translator with a microphone. Members of the audience with questions speak into a microphone, and you listen to the reverse translation through the earphones.

Good translators will have an excellent general technical vocabulary and at least some knowledge in your technical area. They mimic your voice inflections and will read your visual aids to the audience during any pauses in your presentation.

For long briefings, you can expect the translators to work in teams with each handling about a half hour of material before being relieved—because simultaneous translation is very intense work. Once you get used to the equipment, you should be able to just forget about the direct translation process and brief as though the audience were listening to you directly.

9.2.3 Considerations for all translated talks

The following general tips are suggested to help the translator do the best possible job:

- Avoid slang expressions and regional speech mannerisms. ("Y'all" just does not translate very well.)
- Be aware that puns and other types of humor that involve relationships between words in your native language will probably not translate into the other language with the humor intact. To enforce this point, consider the following Spanish language joke as it is literally translated into English:

  Question: How is a train like an orange?
  Answer: Because it does not wait!

  This is definitely not funny in English because the whole humor is contained in the similarity between the sounds of a few Spanish words. In Spanish, the second line is, "Porque no espera." The last two words, "no espera," meaning "does not wait," sound just like the words "no es pera," which means "is not a pear." The second sentence thus has a second meaning, "The train is like an orange in that it is not a pear," which might have been funny had you understood it without several minutes of explanation. If you used that joke, the translator would have had to try to explain it, and perhaps tell the audience why it was funny.
- If possible, use visual aids in the native language of your audience. You will know what they say, even though you cannot read them, and it will make the talk much more interesting for the audience as well as reduce the workload of the translator.
- If your visual aids are not in the audience's language, pause for a moment after presenting each visual frame to give the translator a chance to read it to the audience. Then proceed just as though the audience could read it.
- If at all possible, give your translator (or translation team) a copy of your visual aids, your outline, and any available text well ahead of

time. This will give the translator a chance to practice the special vocabulary for your technical specialty and to look up any words that are not completely familiar.

9.3 Speaking to the nonnative-language speaking audience

Even if the audience speaks your language superbly, the audience will be slower in understanding technical details in your language if it is their *second* language. Effective technical communication with this type of an audience requires that you speak a little more slowly and use better enunciation than you ordinarily do. Be careful to avoid (or explain) slang or regional expressions.

In addition, it is important that you use an adult vocabulary—using oversimplified language would be talking down to the audience. Just slow down enough so that they have time to let the longer words sink in.

9.4 Speaking in a language not your own

If you have some capability in a foreign language, it can be very effective to make your presentation in that language to native speakers of that language. The audience will generally be very impressed by your effort and thus highly receptive to your ideas and very forgiving of any grammatical errors. Still, there are a few cautions:

- You should speak the language at nearly full conversational speed. If you must speak slowly, the audience will tire of the effort of following you.
- Your vocabulary in your subject area should include all of the currently important technical terms.
- Generally, you should be fluent enough so that the information flows. Having a terrible accent and making grammatical errors is okay—the audience will become accustomed to it—but most of the right words need to be there if your presentation is to be effective.

- Your visual aids must be in the audience's language if you are speaking in that language.

Even if you don't speak the language and must use a translator, you can memorize a common greeting phrase with which to open your talk. The audience will generally be charmed by your efforts, increasing the effectiveness of your message. Remember John Kennedy's famous "Ich bin ein Berliner" sentence in the Berlin speech that absolutely captured the audience.

9.5 When something goes wrong

In the real world, things do go wrong occasionally. Projector bulbs burn out at the last minute; room lights fail to dim; people walk in carrying ladders; loud noises start in the next room; and overhead projector slides fall onto the floor, for example. Sooner or later these things happen to everyone. The test of an effective briefer is the way the speaker reacts to them.

The most important thing to remember is to remain calm. Remember that the audience *wants* you to do a good job. Therefore, the audience's first reaction to a small disaster will be to look to see if it is making you uncomfortable. Audience members will be concerned for you because they do not want you to feel bad. If you show that the problem is not getting to you, the audience will relax and wait for you to handle the situation.

Humor is very valuable to remove the audience's discomfort and to show that you are taking the problem calmly. For example, if someone is shouting in the next room, say, "I'm glad he's not mad at us," or "Yeah, that's what I think too." If some equipment fails to work, say something like, "Once again it is proven . . . things hate people!" If you have just goofed (like dropping your slides) say something like, "Mama told me there would be days like this."

Above all, do not just ignore a major distraction. Admit that it is there and fix it. If equipment goes out, get someone to fix it. Call for a short break until the problem is resolved. If possible, rearrange your material to make good use of the audience's time while the problem is being solved. (For example, show the movie after the slide show rather than before it.)

If this is impractical and an alternative visual aid is available (for example, a blackboard), go on as best you can. If your slides are out of order, stop and put them back in order.

When the problem is resolved, thank the audience for its indulgence and carry on. Do not apologize (except for a quick, "Sorry for the delay"), and do not make the audience think that you are really sorry. That would make audience members uncomfortable because they want you to feel comfortable, and they know they are getting the best you have to offer under the circumstances.

9.6 Hostility from the audience

Despite everything said to the contrary in this book, there are circumstances in which someone in your audience may be hostile. Therefore, that person or those people *do not want you to do a good job*. This will normally take the form of heckling, loud ridiculing comments while you are talking, or irrelevant and insistent questions.

The hostility may occur because you are the bearer of bad news; it may be because of prior events that have happened elsewhere (for example, someone has just been chewed out by the boss right before your talk); or it may be that a member of the audience is one of those few people who just like to make trouble.

In any case, the technique for handling hostility is the same: Try to rise above it. Do not let the hostility get inside of you. Give the audience every benefit of the doubt.

Try to defuse the hostility on a personal basis rather than on the basis of your official position. Show that you are sincere in your views and that you have nothing personal against the person or people who seem to be hostile. Do not attack the person who attacks you, because this will tend to make the rest of the audience side with the member who seems to be against you. Rather, gain the sympathy of the audience by showing that you care about that person's opinion even though you might not agree with it.

However, you should also show all members of the audience that you care about their time and are anxious to stop the interruption as quickly and as peacefully as possible. Offer to discuss the matter after the talk. If that fails, ask the person to state a view and then let you answer. In most situations, the audience will side with you and silence the person who is interrupting if you continue to be sincere and follow the guidelines described above.

Some speakers feel that the best way to handle a hostile situation is to disarm it by use of humor. Abraham Lincoln is supposed to have done this on several occasions. Humor can be effective if you are very skillful, but it can also be very tricky in hostile situations. Humor always involves a certain degree of subtlety, and when people are hostile they are prone to misinterpretation of your intentions. If the hostile person feels that the humor is directed against him or her (even though that was not your intention), the hostility will deepen. Sincerity, on the other hand, normally has almost no subtlety to it—so the potential for misinterpretation is far less. Another advantage of sincerity is that it does not take any great skill.

If all else fails, stop your briefing until someone can restore order. If you are the person who needs to deal directly with the problem, do it kindly. Consider this example (actually used by the author in one of those situations), "Bill, I love you like a brother, but you are playing a game while I am trying to do my job. Please let me do it." It worked, and Bill remained a friend.

If order cannot be restored, or if the whole audience is hostile, quit—you are wasting your time. Incidentally, this *almost never* happens.

But what if it is the boss who is hostile? In the unlikely situation that your boss is hostile, the same basic approach will work, except that only *you* can deal with the boss. The audience probably cannot help you in any direct way. The best approach is to stop the briefing and ask the boss directly what the problem is. An excellent opening is, "There seems to be something wrong. Is there something I should know?"

What if the hostile person does not say anything? Sometimes the hostile person will just glare at you with arms tightly folded. Sometimes this will be accompanied by a violent shaking of the head in a "no" gesture. If the person is just glaring, do not look at that person. Look at other people in the audience while you are speaking. If the person is doing something that can be seen by the rest of the audience, you need to handle the situation just as you would if the person were heckling out loud, except that you have to start by asking if something is wrong to move the action out into the verbal realm where you can handle it.

10

Managing the Multiple-Speaker Briefing

THIS CHAPTER IS DESIGNED to help you manage a large-scale briefing involving multiple speakers. It focuses on the special challenges associated with planning and managing the efforts of several speakers to meet an overall objective.

10.1 The nature of multiple-speaker briefings

Multiple-speaker briefings most commonly take two forms. The first is an in-house effort in which a number of individuals from the same organization present a large-scale briefing to explain something to a group of clients, the organization's management, a review committee, or a group of people who must be brought up to speed. The second is a conference in which briefers from different organizations come together to share knowledge with a large, technically qualified audience. Some of the material in this chapter is more appropriate to one form than the other, but all of it should be helpful to you when you are responsible for management of a multiple-speaker briefing of either type.

The top-level structure for an effective multiple-speaker briefing is very much like that for an effective single-speaker briefing. It has an introduction, a body, and a conclusion. It is designed to meet a stated objective that involves some sort of action from the audience. The introduction and conclusion, even though they are complete talks rather than just parts of a single talk, still have the same functions.

The primary difference between the multiple-speaker briefing and a single-speaker briefing is typically scope. In the longer duration of a multiple-speaker briefing, subjects of significantly larger scope can be covered than would be appropriate to a single-speaker briefing. In addition, when there are multiple speakers, it is possible to have individuals with expertise in different fields cover the parts of the subject in which they have special qualifications.

Still, to be effective, the multiple-speaker briefing should not be just a series of unrelated talks—even if they happen to be in the same subject area. In such a briefing, speakers will have overlapping material, assume that someone else has covered something, and otherwise compete with as much as complement one another.

The effective multiple-speaker briefing is planned as carefully as an overall briefing. Then the individual briefings are assigned parts to fulfill in the top-level plan for the overall briefing. The multiple-speaker briefing will have an introduction (opening remarks by the meeting manager), a conclusion (closing remarks by the meeting manager), and a body consisting of individual talks by different speakers. The functions of the introduction, body, and conclusion are the same as for the individual talk.

To be effective, the multiple-speaker briefing must be designed to meet an objective that is clearly understood by the participants. It is best to think of the multiple-speaker briefing on two levels: Each individual talk has a purpose, and that purpose supports the purpose of the larger-scale talk.

10.2 The briefing plan

10.2.1 Set objectives

The planning of a large-scale, multiple-speaker briefing starts the same way as the planning for a single-speaker briefing—by deciding why you are giving the briefing. This should be stated in terms of some action to be taken by the audience.

10.2.2 What you want remembered

The second step is also the same as that for the individual briefing: Decide what very few things you want the audience to remember. This should not be as restrictive as the two or three items that an individual speaker will try to have people remember, but the number of items you expect people to mull over in their minds as they leave should still be a small one.

10.2.3 Scope the briefing

As in the individual briefing, it is important to be sure that you scope the briefing properly. The scope starts with the goals for the briefing. The material should be adequate to achieve the goals of the briefing, and no information that does not contribute to the achievement of the goals should be included.

Another constraint on the scope of the large-scale briefing is the time available. If a half day, a day, or two days are all that are available, the scope (and the objectives, if necessary) must be adjusted to fit the time available.

10.2.4 Plan the flow of information

In planning the flow of information, it is best to start by making a top-level outline for the whole talk. Then, decide on the subjects to be covered by individual briefings and decide on the optimum order for the briefings to make the flow of information most logical to the audience. Pay particular attention to subjects that must be covered as prerequisite background for other subjects. When dividing the subject matter into individual briefings, try to make the amount of material appropriate for cohesive 30- to

60-minute briefings, since these are the most commonly used time blocks for briefings.

10.2.5 Allocate the time

To start, make a rough schedule covering the full period of the series of briefings. The schedule should show starting and ending times for each day, lunch breaks, coffee breaks, and blocks of time for briefings.

While making this schedule, many briefing managers try to fit too much information into the day. Do not succumb to the temptation! People listen and absorb better if they have an adequate number of coffee breaks and lunch breaks. Remember the acronym mentioned earlier: BABE for "The Brain can Absorb only what the Backside can Endure." Too much material in a day will only cause "more wind to blow." It will probably diminish the amount of knowledge that is transferred.

Try to estimate the amount of time it will take to cover each subject adequately. This is, of course, interactive with the information flow design.

Next, fit the outline of the full briefing into the schedule you have just made. If possible, group the individual briefings into blocks of similar subject matter that fit neatly into the blocks of time between lunches and coffee breaks.

10.3 Choosing the team

Be sure that you start this process by considering what qualifications a speaker for each of the individual briefings should have. Then, try to make the best fit with the resources you have available. Doing it in this order will allow you to take the most objective look at how the material should be divided. It may also tell you that you need to change the scope of the briefing because an appropriate briefer is not available.

There are two schools of thought on the assignment of people to give briefings. One school says, "Pick the person who knows the most about the subject, even though that person may not be a particularly good speaker." The other school says, "Pick the best speaker, even though that person may not personally know much about the subject."

There are merits to both approaches. If the experts are chosen, they can be expected to have the most accurate information and will certainly be able to do the best job of answering questions from the audience. However, if an expert is a poor speaker (and some of the smartest people are notoriously bad speakers), it will be difficult for the audience to get the information that the speaker has to offer.

If good speakers who have little or no direct, personally earned knowledge about the subject are chosen, there is a distinct danger that they may say something that is incorrect. Also, they will not be able to answer questions from the audience adequately.

This is one of those difficult decisions for which managers earn their salaries. The solution favored by many experienced briefers (including the author) is to lean in the direction of the speaker who knows the subject well for the transfer of hard technical data but to use a talented speaker to deliver material in a situation where you are trying to make a sale. The person who will be making the opening and closing remarks and speaker introductions should be a good speaker.

The question naturally arises, "Why not teach the poor speakers to speak well and teach the uninformed speakers the subject matter?" The answer is, "Great, if you have the time." As a practical matter, you can certainly help the poor-speaking, technical expert do a better job by asking for "goals" and "two or three things to be remembered by the audience," by reviewing the visual aids for readability, and by giving suggestions during the rehearsals. Also, you can provide support to the uninformed but smooth-talking speaker by having the experts sit in on rehearsals and stand by in the audience to answer any difficult questions that are asked.

10.4 Maintaining unity

Maintaining unity in a series of briefings is an art. This section will present some suggestions, but in the final analysis you must sit back and go through the material to determine if the whole set of briefings works well together. Will the percentage of time spent on each subject seem appropriate to its relative importance and complexity? Do the visual aids in the different individual briefings look like they are designed to be used together? Will the audience move comfortably through the subject matter and come away with a coherent story?

10.4.1 Balance

Some subjects require more explanation than others. By giving the more complex subjects more time than the simpler subjects, you will be allowing the audience time to absorb material at about the same rate. Another way to achieve this effect is by dividing complex subjects into two or more briefings, so that, although each briefing lasts the same amount of time, the rate of information absorption by the audience is fairly even.

10.4.2 Visual aids

If the multiple-speaker briefing truly aims to tell a single, large-scale story, it is an excellent idea to have a common header on all visual aids and to use consistent background colors against which the information is presented. The most common practice to promote unity in visual aids is to require that all speakers use a single type of medium—usually either overhead projector slides, 35-mm slides, or VGA projector. All are required to use horizontal layout for the slides, and all are required to use a specified header. The header, which is about 10% wide across the top of each frame, is a field of a specified color that includes an organizational logo in a fixed position at one side and that may include a logo for the project being briefed at the other side. Between the two logos, there is space for the title of the slide. The body of the slide contains the rest of the information. Figure 10.1 shows an example of this type of standard layout.

If you are using direct computer projection, there is a temptation to put all of the briefings on the same large disk or CD so that no one will lose his or her slides and so that there are smooth transitions. In general, this is a bad idea because it restricts your ability to reorganize your briefing on the spot to overcome some challenge. Schedule conflicts may restrict the availability of critical individuals in the audience, dictating a change in speaker order. There are a hundred other reasons that something may need to be changed at the last minute, but you get the idea.

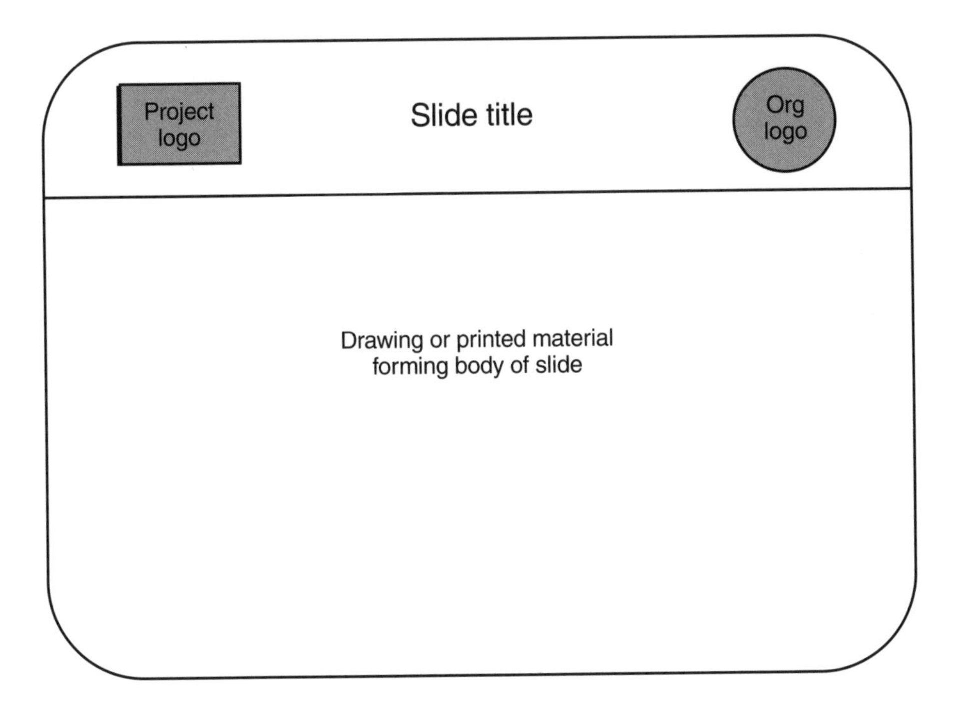

Figure 10.1 Example of required slide format.

10.4.3 Sequence of briefings

In the process of getting an education, every technical professional has taken a class without having taken a prerequisite. The result was that the student spent the whole semester trying to figure out what was going on. This can happen, to a lesser degree, in a series of briefings in which one speaker fails to build on the foundation laid by another speaker.

The manager of the series of briefings should review the top-level outline to be sure that the order of the briefings supports the unity of the series. A few effective organizational approaches to maintaining unity in the order of briefings are listed as follows:

- Tutorial matter first, followed by the meat of the briefing;
- Subject matter organized from general to specific;
- Moving from background to technology to application;

- Moving from past to present to future;
- Moving in the direction of process flow.

10.5 Management considerations

The manager of a multiple-speaker presentation needs to manage the schedule, the product, and the process flow.

10.5.1 Preparation schedule

Like any job that requires the efforts of several people, the multiple-speaker briefing must be managed with intermediate milestones. A preparation schedule should be generated and distributed very early in the process. Items to be shown in the schedule include, but are not limited to, the following:

- Date by which individual briefing outlines must be submitted;
- Date by which visual aid rough drafts must be into the production/graphic arts department (unless speakers are to prepare their own final visuals);
- Date by which final visual aids must be ready;
- Date by which handout material must be ready for printing;
- Date for rehearsal;
- Dates of the briefing.

10.5.2 Schedule of the briefing itself

Once you have prepared a briefing schedule, half of the work is done. The other half is to stay on that schedule during the day or days of the briefing. Very often, a speaker will become enthralled with the material and simply not notice when it is time to quit. Sometimes, if left unchecked, a speaker will speak twice as long as scheduled. However, a much more common problem is that each speaker will go just a few minutes overtime. The result, in either case, is that you must shorten or skip coffee breaks and may even have to skip the last speaker to finish in time for people to catch their planes home.

The only solution is to keep every speaker within the allotted time. Explain to all of the speakers ahead of time that the schedule will be strictly enforced and then remove the first violator from the stage at the end of the allotted time.

This may seem harsh and unprofessional, but it is a fact that audiences become extremely impatient when they have been led to expect a talk to last a certain period of time and it goes beyond that. Even the most interesting speaker will lose significant audience involvement as soon as the talk starts to go overtime.

A way to manage the time with more subtlety is to use a timing light that has green, yellow, and red lights and perhaps a flashing red light. Arrange the position of the light so that the speaker can see it but the audience cannot and have someone with a stopwatch run the light. Tell each speaker the following:

- The green light is available for any sort of internal time marker desired.
- The yellow light will come on when the speaker has some precise amount of time remaining (perhaps 10 minutes).
- The red light (or flashing red light) will come on when the speaker has five minutes to go.
- The speaker will be removed from the stage, bodily if necessary, five minutes later. (If the speakers are bigger than you, smile when you say this last one.)

The result is that speakers will clearly see the approaching end and have a chance to wrap up the presentation gracefully and get off the stage on time. This technique really does work. In one national-level, industry briefing, an army major general famous for his filibusters was heard to say, "Well, I see that the %#*@! light is blinking so I better quit."

Another scheduling consideration is when out-of-town speakers must catch their planes home. This must be considered as soon as speakers are selected, so that any speakers who must travel can make appropriate plans. If you can do so without effecting the quality of the series of briefings, it is a good idea to schedule out-of-town speakers early in the day to preclude a last-minute problem with plane schedules.

10.5.3 Visual-aid equipment

Have each speaker specify well ahead of time the type of visual-aid equipment that he or she needs and have it available in the briefing room. If you have some flexibility in the order in which individual briefings are given, try to schedule any talks requiring extensive equipment setups right after a break, so that the equipment can be set up before the audience returns.

10.5.4 Handling slides

Be sure that you have a carefully considered plan for getting visual aids from the speakers and returning them. Whatever the plan, it should be communicated to the speakers well ahead of time and followed faithfully during the briefing itself.

It is a good idea to make each speaker responsible for getting the slides to a coordinator at the beginning of the day on which the talk is scheduled. Then, have each speaker pick up the slides during the first break after his or her briefing.

10.5.5 The hot seat

To help your series of briefings flow smoothly, you may want all of your speakers to sit in a particular part of the room so they can be found easily. Whether this is done or not, it is an excellent idea to have a "hot seat" in which the next speaker sits while the previous speaker is presenting. The hot seat should be located very near the point at which speakers will mount the stage or approach the lectern.

10.5.6 A speakers' breakfast

Many experienced briefing managers like to have a speakers' breakfast on each day of a multiple-speaker briefing. This gives the meeting manager a chance to explain the schedule and procedures (for example, pointers, switch locations, and the timing light procedure) to all of the speakers. The speakers' breakfast also causes the speakers to arrive a little early, preventing the normal tenseness of speakers arriving at the last minute.

10.6 Introducing speakers

Introductions of individual speakers need to fill three functions: to make smooth transitions between subject areas, to establish the speaker's credibility with the audience, and to get the speaker started.

10.6.1 Making a smooth transition

When one speaker has finished, there is a natural discontinuity in the flow of the briefing until the next speaker gets into full swing. The day meeting manager should use the introduction of the next speaker to bridge this discontinuity by dealing with the relationship between the last subject and the next subject to be covered. For example, "Now we move from the outside of the machine to the inside."

10.6.2 Establishing speakers' credibility

There is an "old saw" that defines the purpose of the speaker introduction as answering the question, "Why this speaker to this audience on this subject at this time?" Like most old saws, it is oversimplified but serves well if treated in context. Let us look at the parts of the question:

- *Why this speaker to this audience:* The audience members should be told what this person has done to deserve the honor of speaking to them. The speaker no doubt has academic credentials that will establish the speaker as at least "one of us," if not "one of the most qualified of us." The speaker has probably had some sort of career success to which the audience can relate. Any sort of career accomplishment impressive to the members of the audience will increase their anticipation and their belief that this person is the type of person who is likely to bring some important information.
- *On this subject:* In addition to general qualifications, the speaker will have some special reason to have insight into the subject being

presented. Perhaps the speaker has done some significant work in the subject area.

- *At this time:* The speaker's subject or information should be timely. This is part of the introduction because the subject of the talk probably has special interest to the audience at the present time—because of some world event or technical event that is happening. Timeliness can also come from the fact that the speaker has some new information that has just become available. This is particularly powerful if the audience has been waiting for the information for some time.

The whole effect of this part of the introduction should be to give the speaker the best possible starting point from which to speak. With properly established credentials, the speaker will be able to get away with statements of universally accepted fact that would have to be proven in painful detail by some lesser individual.

10.6.3 Get the speaker started

Many conference meeting managers present some bit of personal information about each speaker to "round" the speaker a little for the audience. Professional credentials tend to be very cold, and a tiny bit of personal information will help the audience to like the speaker as a person even before the briefing. Examples are, "She was a member of a Mt. Everest expedition in 1974," and "In addition to teaching engineering, he is the father of four sons . . . all teenagers."

The end of the introduction should be something like, "*So* please help me welcome Dr. Jones to the lectern." If welcoming applause is appropriate, the chairperson should start it. During the applause, the speaker gets settled in behind the lectern. When the applause dies down, the speaker begins with an opening statement.

Incidentally, it is often very effective to refer to the speaker as "our speaker today" or "he" or "she" during most of the introduction and then use the speaker's name for the first time in the "so please welcome . . ." statement at the end. Even when the audience knows perfectly well whom the introducer is talking about, this practice creates a sense of anticipation that peaks audience interest just at the time the speaker moves to the lectern to start.

10.7 Opening and closing statements

It normally falls to the meeting manager to start and end the full set of briefings with opening and closing statements. Each statement has important functions.

10.7.1 Opening remarks

Even when the first speaker is an important, out-of-town personage who is giving a keynote address to set the tone for the full briefing, it is the duty of the meeting manager to make an opening statement that gives the audience a road map for the day. The opening statement should state clearly the objectives of the full set of briefings. It should also expose the audience to the reasons for ordering the individual briefings as they are.

10.7.2 Closing remarks

The meeting manager's closing remarks should, above all, reiterate the objective of the full set of briefings. In addition, it is normal practice for closing remarks to acknowledge all the people who have participated in the day's activities. However, the acknowledgments must not be too long or too flowery (as they often are) because you do not want to shift the focus of the audience's attention from the objectives of the briefing.

Finally, the closing remarks tell the audience that it is time to get up to leave. A large group of people, unless told to leave, will often sit for an uncomfortably long time waiting to see if something else is going to happen.

11

Real-World Briefings

THIS CHAPTER DISCUSSES the specific requirements for some of the most common types of talks that technical professionals deliver. There is one section for each type of talk, covering special considerations about the audience, the nature of the speaking situation, and special techniques and considerations.

The types of talks covered are the following:

- The work review (including design review and status report);
- A marketing presentation to a major customer;
- A technical conference paper;
- An after-dinner talk to a technical society meeting;
- Reading a prepared paper;
- Convincing the skeptical audience.

11.1 The work review

The work review is likely to be the first real-life, on-the-job speaking task that is required of a technical professional. The stated purpose is to tell a group of people what you have been doing during some period of time, typically a month.

Examples of work reviews include the following:

- Design review (hardware or software) by the designer or design task manager;
- Program review by the program manager;
- Activity report as part of staff meeting.

11.1.1 The audience

The audience can be expected to include some or all of the following individuals:

- The boss, who assigned you the work task by which you have been earning your salary. The boss is responsible for the overall operation, including your task, and is there to find out how well you are sticking to your plan.
- The boss's boss, who is there to see if your boss has the whole operation under control by observing this review of the individual tasks (including yours).
- A customer representative, who is there for the same reasons as the boss's boss.
- Support function representatives, who are responsible for specific aspects of performance (for example, cost analysis, schedule analysis, and safety). They are attending the review to look out for their special interests as affected by your work. (For example, the cost controller is only interested in your budget performance.)
- Your peers, who have tasks of their own to report during the same review. They will be listening to your review so that they will be

aware of any changes in the status of your work that might affect their own tasks.

- Various hangers-on who do not have any real function but seem to come out of the woodwork whenever a series of reviews are being held.

11.1.2 The speaking situation

You have continuing responsibility for some task and need to present a complete and accurate status of your performance of that task to the audience. The audience can be expected to include individuals who are quite knowledgeable in your technical field (your boss and peers) as well as some individuals who are only marginally familiar with the field but are experts in some specialty that applies to your work task. Your boss and peers will be familiar with all of the buzzwords in the field. They will probably be familiar with the overall operation of which your task is a part and with the plan against which you are operating. The other members of the audience will probably not be familiar with your task and have only marginal knowledge of the overall operation, but they will know a great deal about the general requirements and terminology associated with the specialties for which they have responsibility.

Work reviews almost always comprise a series of presentations by all of the individuals who have responsibility for parts of a larger operation. For small operations, the series of presentations may last only two or three hours. For larger operations, the review may involve a number of days of presentations.

The audience can be expected to be alert and in a businesslike mood. They will either be suffering from hours of reviews or looking ahead to the same and so should be expected to be talk-weary. They will be anxious to get the facts with as little extraneous material as possible. The subject is important to the listeners because it directly affects their own job performances.

Visual aid media will normally be specified. Overhead projector slides or VGA projector are the most common media specified for work reviews, although flip charts or 35-mm slides are sometimes used. Another very common practice is to require that a hard copy of visual aids be provided ahead of time to each person in the audience. Sometimes the hard copy is used in place of any other visual aid. Almost always, some unusual visual aid media (film clips, pass around object, etc.) can be used to make an important point.

11.1.3 Presentation format

In many cases, there will be a required format for the work review presentation. This format will normally include the status of budget, schedule, and performance aspects of your work. Whether or not required, it is a very good idea to feature problem areas as a separate item—particularly those that require some action by the boss to allow you to overcome the difficulty. Perhaps you need additional budget or a schedule variance because of some unexpected difficulty.

If no format is specified, the following sample format is appropriate for most work reviews:

1. Introduction:
   - Identify the organization or activity for which report is made.
   - State the scope of task or responsibility.
   - Identify your relationship to the task (such as manager of group).
2. Technical description of work done (if required):
   - Describe the design if this is a design review.
   - Describe changes to the design if this is a design modification review.
   - Describe the responsibility of organization if required.
3. Milestones achieved during reporting period:
   - State the schedule milestones met.
   - Mention units of work completed (for example, cleaned out 572 horse stalls).
   - Mention significant staffing changes.
   - Recite previously outstanding problems resolved.

4. Status summary:
   - Give the progress level on assigned tasks.
   - Give the schedule status.
   - Give the budget status.
5. Problem areas:
   - List the problem areas.
   - Include one visual on each problem: statement of the problem, statement of alternate solutions considered, statement of suggested solution, and required action by boss (typically, authority to proceed).

11.1.4 Special techniques and considerations

An important element of this kind of technical presentation is to establish and maintain your credibility. You are giving the briefing because you are an expert in some field, responsible for some important activity, or both. The special vocabulary of your profession is important to this credibility, so use it during the talk. However, it is also important to have your audience know what you are talking about, so be very careful to define your terms. Consider the following:

- Many in your audience are likely to be in high positions but have limited knowledge of your specialty. Be very careful to show respect for them (and not patronize them) when you are explaining "simple" things that they do not understand; many technical professionals do not understand aspects of your work. Ways to do this include using nontechnical but college-level vocabulary to define the special terms in your field, using graphic explanations (including sketches) rather than complex math, and explaining the physical significance of any math expressions that must be used. In the movie, *Coal Miner's Daughter,* Loretta Lynn is quoted as saying, "I'm just ignorant, not stupid!" Someday, when you are a high-level manager listening to reviews like this, you will feel that way too.
- Do not be afraid to use a special visual aid for any good reason. Remember that this is a talk-weary audience. They will appreciate a

change of pace if it is appropriate. The best examples of special visuals for this circumstance are an actual piece of equipment being discussed (particularly powerful if it is something small that can be passed around) or a videotape or film to show some activity that has to be seen in action to be understood.

11.2 The marketing presentation

Sooner or later, you are going to have to help sell something. This involves the preparation and presentation of a technical marketing briefing to motivate a potential customer to buy a product or service offered for sale. The same techniques apply if you are selling an idea to the management of your organization or to your colleagues, so this type of talk should be mastered by more than just marketers.

When selling soap to the general public, a company uses advertising professionals who are expert in dealing with a mass audience who make relatively casual decisions to buy one product over another. However, when a company wants to make single sales of costly equipment to sophisticated customers, technical professionals capable of understanding what makes the product better than "brand X" are required to participate directly in the selling process.

Often the first contact that a technical professional has with the customer is the presentation of a marketing briefing. This talk is important, not only because it might lead to a sale, but also because it often entails very high visibility within the customer community and within the briefer's own organization. A well-presented marketing briefing has put many a career on the fast track.

11.2.1 The audience

Above all, the audience will include the decision maker who has the authority to buy what you want to sell. The decision maker will typically be someone who may have had expertise in some technical area in the past but has been in a management function too long to be aware of the latest developments in the field. Decision makers *do* understand the fine points of money.

The decision maker may be supported by experts who are capable of understanding any level of technical detail you can present. However, you must remember that it is the decision maker who will make the decision. The audience typically includes someone from your management and marketing representatives from your organization, who will know the general field but will know less than you do about your technical specialty. The total audience size for a marketing presentation is typically small, perhaps 6 to 12 people.

11.2.2 The speaking situation

The marketing presentation is usually a stand-alone briefing. The audience has been gathered specifically for this presentation and will leave afterward. The briefing will normally be held in a large office or small conference room with the members of the audience seated around a table so that they can take notes conveniently. You cannot rely on visual aid equipment being present. An overhead projector is sometimes available if you arrange for it ahead of time, but even this is not definite. If you require any other visual aid equipment (such as a videotape player or 35-mm slide projector), you will probably have to carry it in under your arm. Those new, lightweight VGA projectors are ideal.

If your marketing briefing includes a demonstration, be careful of the logistics. In an office or conference room setting, standard electrical power will almost always be available. Since standard power varies widely from country to country (for example, 117 volts, 60 Hz; or 220 volts, 50 Hz), it is very important to be sure that your equipment will work with the local power if you are demonstrating in a foreign country. This seems obvious, but many an expensive demonstration far from home has failed to take place because of this simple mistake.

It is very dangerous to assume that anything but standard electrical power will be available at the site of the demonstration. You should take everything you need with you or make very specific arrangements to have it available for your use when you arrive.

11.2.3 Presentation format

Since the sale will presumably make the customer better off after the purchase than before, the "background, problem, solution" format is often used. ("You have a problem, and my product will make it go away.") However, almost any type of format can be appropriate for the marketing presentation, depending on the nature of the product or service being sold.

It is strongly suggested that the "two or three things the audience remembers" (see Chapter 2) be the two or three strongest selling points for the item you want to sell. The one, absolutely essential format item is that you ask for the sale as your call to action in the conclusion.

11.2.4 Special techniques and considerations

Features versus benefits

The classic mistake made by technical people in a marketing briefing is that they focus on the features of the product or service offered for sale. The customer does not make a buy decision based on the features of the object of the sale. The decision to buy is based on the benefits that the customer expects to receive as a result of having made the purchase.

To sell anything—from magazine subscriptions to airliners—you need to put on your "customer hat." Look at the sale from the customer's point of view. How will the customer be happier by having decided to make the purchase you are suggesting—will it lead to reduced costs or increased profit, for example? When you describe features, design processes, or manufacturing processes, do so only to help the customer believe that the benefits will really happen.

Money matters

As mentioned earlier, decision makers understand money. Be sure that you cover all appropriate aspects of the subject that are related to money. Many buy decisions are made not on the immediate cost of acquisition, but on the long-term cost of ownership (e.g., maintenance). If the object of the briefing is intended to increase the customer's profits, be sure that the story hangs together financially.

Columns of numbers

Managers understand numbers. Be sure that your numbers add up. It is often necessary to deal with dissimilar sets of numbers on a single visual aid. Examples include the following:

- Input quantities that add to subtotals;
- Physical quantities and the resulting costs.

It is important that the roles of all numbers on visual aids be obvious. If you show anything that looks like a column of numbers, the managers in the audience will try to add them up to see if your total is correct. Checking numbers is one of the ways that good managers use their general wisdom about people to determine whether or not a technical person is on top of the subject.

Technical level

The technical level must be deep enough to give the decision maker confidence that you know what you are talking about. However, it must be no deeper than necessary.

If the decision maker does not personally have expertise in the technical field, you can be sure that the technical support people the decision maker brings along will detect any technical inaccuracy in your presentation.

Bad news

Since this is a sales pitch, you will, naturally, focus on the strong points of the product or service you are selling. However, no product is perfect, and every product has some sort of competition. Your customer probably has some knowledge of the weaknesses of your product and almost certainly knows about the competition.

Good selling practice requires that you deal with the bad news in a positive way. Show that the shortcoming does not affect the customer in the anticipated application, or show that the other virtues of the product offset the shortcoming (for example, correction of the problem would double the price).

Deal with the comparison of your product with its competition. You may not want to point out that the competition exists, but be sure that your talk includes coverage of the areas in which your product is better.

Notes made by the audience

The audience will probably be seated around a table and will be taking notes while you are speaking. One measure of a marketing presentation is what the audience is writing down while you are speaking. Ideally, those notes should include all of the reasons the sale should occur.

You cannot directly move the audience's pencils, but your selection of visual aid formats and the structure of your introduction and conclusion will strongly effect what is written down. Handout material covering detailed information in the way you want it remembered is also very effective.

11.3 The technical conference paper

Presentation of a technical conference paper is one of the most prestigious of activities in which a technical professional participates. It is the way to share the results of research or other significant professional activity with peers. In some professions, the presentation of such papers is an absolute requirement for career success. In other professions, it is an optional activity, but acceptance of an invitation to present a conference paper is almost always a good career move because of the exposure it creates.

11.3.1 The audience

The typical technical conference audience is alert, knowledgeable, and highly motivated to hear your paper. It can be expected to be made up largely of experts in the general technical field for which the conference is designed.

Since conferences are often organized with parallel presentations of papers, the attendees may have to choose which papers to hear. Even when there is no time overlap for presentations, few attendees listen to every paper. This means that the audience listening to your presentation has specifically chosen to do so and will tend to have a high level of interest and background information in your area of expertise. They will not be bashful about asking hard questions.

The listeners can be expected to be alert because they have come to learn something new and important in a field that is important to them. They consider listening to your paper as important work. They came to learn, not to be entertained.

11.3.2 The speaking situation

First, you can expect to be one of many presenters in a series of briefings that will last multiple days. You will probably be asked to use 35-mm slides, VGA projection, or overhead projector slides for your visual aids, but you can usually request other visual aid devices if you do so well ahead of the conference.

There will usually be a "proceedings" available at the conference, containing either abstracts of the papers to be presented or actual copies of the papers. If the audience has copies of your paper ahead of time, you should use the techniques for this situation described in Section 11.5.1.

You can expect a question-and-answer period immediately following your presentation. Informal questioning will likely take place during coffee breaks by individuals with particularly high interest.

11.3.3 Special techniques and considerations

The following considerations apply to all technical paper presentations:

- Accuracy is extremely important in a technical paper. It is worth the trouble to double check all of the information presented and to have someone else independently check your work. Are the experts properly quoted and credited? Are the facts well established? Do all numbers add up?
- When you are dealing with subjects in which there is open controversy, carefully identify and admit the controversial nature of any background material and conclusions that fall outside of the realm of facts that most can accept. Some of your audience can be expected to fall into camps on each side of every issue, so the techniques in Section 11.6 on convincing the skeptical audience will be valuable.
- Technical conference papers often aim to remove some item from the realm of controversy and to place it into the realm of established fact. If this is accomplished, it is a historically important event to your technical field, and the paper should be focused tightly on that goal.

- Because technical conference papers are so highly focused on the significance of the ideas they contain, the presenters often do not spend any effort on the development of effective presentations. Some just stand up and read their papers. The result is that conference papers are often deadly dull. If you take the time to make the presentation of your technical conference paper an effective briefing, just as you would if it were a marketing presentation, the audience will remember you and your ideas with extra warmth because your briefing will be such a comfortable relief from most of the other papers.

11.4 The after-dinner technical talk

There are many societies serving technical professionals in various fields. Most of these societies have periodic meetings, usually dinner meetings with speakers. Few of the local groups in these societies can afford to hire professional speakers. What this means is that there is an inexhaustible supply of public speaking opportunities for technical professionals who are willing to "sing for their supper." Speaking to a professional society meeting is an excellent way to gain recognition and stature among peers.

11.4.1 The audience

Audiences for this type of talk are normally homogeneous. They tend to have approximately the same education level and to know the jargon of the profession. They probably get much of their opinion about the professional field from a society magazine or journal. They will probably be fairly close to each other in socioeconomic level, political opinions, and other "prejudices."

11.4.2 The speaking situation

Please reread Section 9.1 about speaking in a restaurant situation. Although this is a talk to a group that is very well versed in the subject matter, remember that they have just eaten and are in a restaurant situation. The talk should be designed accordingly.

The audience will expect to learn something from the talk, even though they are not particularly predisposed to learning after eating. Therefore, it is important to say something important but to limit the subject and to present the subject matter in an entertaining way.

11.4.3 Special techniques and considerations

The after-dinner talk requires some special considerations.

- Your credentials as a speaker are particularly important in this kind of a talk. The audience expects to hear someone who "is worth listening to." Your introduction should show clearly why you can be expected to have some special knowledge of interest to the group.
- Show that you are "one of us." In the introduction of your talk, establish rapport with the audience. You either have the basic credentials (for example, you have a medical degree if you are speaking to a group of M.D.s)—or you have worked with the profession represented by the audience and understand on a deep level how they work and what their problems are.
- Use visual aids that can be seen and are effective in the restaurant situation. Unlike other types of restaurant situations, the group may have taken the trouble to establish a regular meeting place in which a wide range of visual aid media will work.

11.5 Presenting a prepared paper

In many of the more prestigious professional speaking situations, the speaker is expected to prepare a formal paper and present it to the audience. To the speaker who wants to do a good job, this is a very challenging type of talk.

Two cases are considered here. In one case, there is a list of proceedings for the conference, and the audience has a copy of the paper you will present. In the second case the audience does not have access to the paper, so you must read it to them.

11.5.1 Audience has copy of paper

If the audience has a copy of the paper you will be presenting, you need to assume that they have read it. Unfortunately, not all of them will have read it beforehand, some will never read it, and some will be reading it for the first time while you are speaking. Still, the members of the audience who are most interested in the subject will have read it and will expect to receive some additional information from your presentation.

Above all, do not read your paper to the audience. This would bore them to tears and would be a complete waste of your time. Rather, your talk should be a presentation of the basic information in the paper with additional bits of background information and examples to make it more memorable.

To be effective, your talk should do the following:

- Have visual aids based on the figures in the published paper;
- Feature the introduction and conclusions from your paper;
- Have as the main points of the body the major sections of the paper;
- Emphasize the two or three things you want the audience to remember (as described in Chapter 2);
- Cover the same scope as the published paper.

11.5.2 Audience does not have paper

If you are to present a prepared paper and the audience does not have a copy of the paper before your presentation, you will be expected to read it to the audience. An important special case is the "third-party" presentation of a paper written by some famous but absent person. This kind of presentation is challenging because it is difficult to read your own or another person's paper in an interesting way.

Many presenters will keep their heads down and simply read the paper aloud in a monotone voice, which is painfully dull to the audience. However, a well-read paper can be quite interesting, particularly if its contents are interesting to the audience.

Some tips on how to read a paper aloud effectively follow:

- Make visual aids from all of the figures in the paper.
- Have someone follow along with your reading on a specially marked copy of the paper that shows when to put up each visual aid.
- If possible, have a special copy of the paper made for you with large type. This is easy if the paper is available in a computer that accommodates different type fonts.
- Read the paper several times before your presentation so that you will be completely familiar with it.
- If someone else wrote the paper, be sure that you can properly pronounce all of the words used.
- Highlight key words in the text so that you can see at a glance where you are in the text after you have looked up at the audience.
- Practice reading the talk from the actual copy you will be using during the presentation.
- Try to read the paper with emphasis and conviction. Vary your rate, pitch, and volume (see Chapter 8).
- At the beginning of your presentation, give a few words of personal introduction speaking directly to the audience before you start to read the paper. When someone else has written the paper, the introduction should be from *you* (for example, "Dr. Jones was very disappointed that he could not be here to present this paper himself, as he was called away to quell a riot at the North Pole. I worked with him in the development of the research described, and I am honored that he chose me to present this paper for him.").
- Make as much eye contact with the audience as possible while you are reading. If you are very familiar with the paper, you should be looking at the audience most of the time.

11.6 Convincing the skeptical audience

It is easy to convince an audience that already agrees with you, but sometimes the goal of your presentation is to try to change their opinion. When trying to convince a skeptical audience, you are selling ideas, but you have the additional challenge that they do not believe that they want to buy what you have to sell.

It is important to approach this kind of talk with a clear understanding that you are dealing with a difference of opinion, not personal enmity. Be sincere in your presentation and let your whole bearing show that you are dealing with technical issues, not with personality considerations.

The following tips should help you prepare to convince a skeptical audience:

- Try to avoid an adversarial relationship with the audience. Rather, try to establish a position that you and the audience are both looking at the subject from the same point of view—with an open and inquiring mind.
- Avoid, as much as possible, stating your personal opinions. Let the disagreement be between the audience and the facts rather than between the audience and you.
- Include as much objective information as possible (including test data, and opinions from established experts).
- Build on the opinions that the audience already holds.
- Show that you respect the opinions of the audience.
- Limit your objectives as much as possible. It is better to succeed in changing the opinions of the audience by a small amount than to fail in moving them a long way.

A

Speaking Exercises

THIS APPENDIX CONTAINS seven public-speaking exercises. If you are not using this book as part of a class (with a hard-nosed teacher who would make you do these in class), consider doing them anyway. No amount of theory will, by itself, make you a good speaker—you *must* practice. By preparing these talks and delivering them to a mirror and a tape recorder (family dogs also make excellent private audiences), you will learn a great deal with a moderate investment of time. Also, it is surprising how many opportunities arise to present a talk that you "just happen to have ready."

Each exercise covers a different technical public-speaking situation and includes an explanation of the assignment and an evaluation guide to help you determine how well you have completed the assignment. Each exercise also references the sections of the book that give background information appropriate to that particular type of technical talk.

If this book is being used as a text for a course, the instructor or another student should provide a written and oral evaluation (against the evaluation guide) of each exercise for each student. If the written evaluation is entered in the speaker's book in real time while the student is speaking, it will create a permanent record that will help the student gain the most from the assignment. Appendix C contains a set of evaluation forms for all the exercises.

Exercise 1: the work review

The speaking project

Prepare and present a 10-minute, work review briefing on your current work over the last calendar month. This can be a project or task status review or a review of your organization's activities. Assume that you are the manager of the task or the organization being reported and that your talk is one of several being presented in the same half-day meeting. The audience includes all of the managers of the other tasks or organizations in the whole program or larger organization. It also includes your boss, your boss's boss, and representatives of every support organization associated with your work. Review Section 11.1 for background information on work review presentations.

Evaluation guide for the work review

1. What action did the speaker apparently want the audience to take?
2. What will the audience remember two hours after the talk?
3. Was the discussion limited only to material required to fulfill the objective?
4. Were a clear introduction, body, and conclusion present?
5. Was the speaker knowledgeable about the material, and did he or she organize and present it effectively?
6. Did the speaker properly define and use the vocabulary of the profession?
7. Did the speaker properly explain technical concepts without patronizing the audience?

Exercise 2: the marketing presentation

The speaking project

Prepare and present a 10-minute marketing presentation on any product or service relative to your work. It is not necessary that the product or service be appropriate for sale to the general public. It can be something highly specialized, like a modification to a major system that a single customer owns. You might choose to "sell" a departmental reorganization to your boss. This is acceptable because the boss must make a single "buy" decision at the time that official approval of the departmental organization is made.

There are two important characteristics of this project. The first is that you attempt to motivate one or more potential customers to make a decision to buy something. This is not just a talk to convince someone of a point of view—it must be directed toward a single sale. The second is that the presentation be a technical briefing to support the sale—not a television commercial to sell soap. The talk should be directed specifically toward your audience. The presumption is that there are relatively few immediate customers for this product or service and that a technical understanding of its nature or usefulness is important to the sale. Review Section 11.2 for background information on the marketing presentation.

Evaluation guide for the marketing presentation

1. Did the speaker make the sale? Why, or why not?
2. Did the speaker focus on benefits rather than features?
3. Were the technical detail level and vocabulary of the talk appropriate?
4. Were the visual aids effective? Could they be easily read?
5. Were numbers properly presented?
6. Were shortcomings and competition dealt with in a positive manner?
7. Look at the notes you made during the talk. Do they support the sale?

Exercise 3: the technical conference paper

The speaking project

Prepare and present a technical conference paper in your professional field. The specific subject can be either your current work or some project you have recently completed. The conference will have papers scheduled every 30 minutes, so your presentation should be 20–25 minutes long to allow time for questions from the audience. The conference has no proceedings, so you do not need to write a formal paper for submission. Review Section 11.3 for background information on the conference paper presentation.

Evaluation guide for the technical conference paper

1. Was the information in the paper completely accurate?
2. Did the speaker have proper backup for controversial subject areas?
3. Were the technical detail level and vocabulary of the talk appropriate to a sophisticated audience?
4. Were the visual aids effective? Could they be easily read?
5. Was the tone of the paper appropriate to someone working at the state-of-the-art level in the field?
6. What three things do you remember from the talk? Do they support the speaker's objective?

Exercise 4: the after-dinner technical talk

The speaking project

Prepare and present a technical after-dinner talk to a professional society. You will be the keynote speaker at a dinner meeting of a professional society in your technical field. The schedule calls for a short social hour with a cash bar, then dinner, then a short business meeting, and finally your talk. It is customary for speakers to make a formal presentation lasting 20–25 minutes and then to answer questions for 15–20 minutes. The whole meeting will take place in a private room in a restaurant. A screen is available, and the lights can be dimmed, so you can use projected visuals if you wish. Review Sections 9.1 and 11.4 for background information about the after-dinner technical talk.

Evaluation guide for the after-dinner technical talk

1. Was the presentation both informative and entertaining?
2. Did the speaker avoid controversy?
3. Were the technical detail level and vocabulary of the talk appropriate to a technically qualified audience that has just finished dinner?
4. Were the visual aids effective? Could they be easily read in the restaurant situation?
5. Did the speaker establish himself or herself as an "insider" to the audience's professional orientation?
6. What three things do you remember from the talk? Do they support the speaker's objective?

Exercise 5: the multiple-speaker briefing

The speaking project

Plan a multiple-speaker briefing in which a complex subject will be covered in a full day of briefings, present your opening and closing remarks, and introduce the speakers. The project has the following steps:

- Pick a subject appropriate to a full day of briefings (for example, if you are a software professional, you might use a full-day tutorial on artificial intelligence).
- Select the speakers (go ahead, use world famous experts if you like).
- Plan the complete set of briefings, setting the scope and objectives for each speaker and the schedule for the day.

- Make a written set of guidelines for the speakers.
- Plan and present your opening and closing remarks and the introductions for each of the speakers.

Review Chapter 10 for background information about the multiple-speaker briefing.

Evaluation guide for the multiple-speaker briefing

1. What was the overall objective for the set of briefings? Was the objective met?
2. Was the subject area divided properly into individual papers?
3. Did the opening remarks set the proper tone for the day and let the audience know what to expect?
4. Did the closing remarks summarize the main points made during the day and focus the audience's attention on the overall objective of the day of briefings?
5. Were the written guidelines to the speakers adequate and appropriate?
6. Were the introductions effective? Did they establish the speakers' credentials?

Exercise 6: reading a prepared paper

The speaking project

The object of this project is to read effectively a prepared technical paper to the audience. You should pick a paper from the proceedings of a recent technical conference in your professional field. Study the paper carefully and make visual aids from its figures. If additional visual aids are required for an effective presentation, generate them.

Work out a way to have the visual aids displayed at the appropriate times during your reading of the paper. Suggestions are to have someone else change overhead projector slides according to a keyed text, to have someone else change visual aids in response to your verbal request, or to use a remotely controlled 35-mm slide projector or VGA projector and change the slides at the appropriate times.

Read the paper to the audience in an interesting and effective way, making proper use of visual aids. Review Section 11.5 for background information on reading a prepared paper.

Evaluation guide for reading a prepared paper

1. Was the paper, as read by the speaker, interesting to the audience?
2. Did the speaker achieve good eye contact while reading the paper?
3. Was the speaker able to read the paper so that it sounded spontaneous?
4. Were the visual aids effective? Could they be easily read?
5. Were the visual aids appropriately displayed during the reading of the paper?

Exercise 7: convincing the skeptical audience

The speaking project

Prepare and present a 15-minute talk to convince a skeptical audience to accept a technical conclusion that is contrary to their present beliefs. This is not a talk to a hostile audience. It is a talk to deal with a technical difference of opinion and does not concern emotionally held views.

For this assignment, pick a subject area in your field of technical expertise in which there has been a significant breakthrough that changed the way the profession does something. Assume that you are speaking to professionals who have not heard about the breakthrough and are convinced that the old way is the only way. In your talk, convince the audience with objective facts that the new way will really work. For example, if you are an aircraft engineer you might want to convince a group of professionals, in related fields, that an aircraft with a swept-forward wing will really fly. Review Section 11.6 for background on the talk to convince the skeptical audience.

Evaluation guide for convincing the skeptical audience

1. Was the audience convinced?
2. Did the speaker have convincing, objective backup for all assertions contrary to the audience's previously held views?
3. Were the technical detail level and vocabulary of the talk appropriate to the audience?
4. Were the visual aids effective? Could they be easily read?
5. Was the author able to keep the controversy on an objective, technical level without being personally threatening to the audience?
6. What three things do you remember from the talk? Do they support the speaker's objective?

B

Speech Notes: How and When To Use Them*

Abstract—The generation of speech notes should be an integral part of the preparation of a speech. Such "notes" can be actual notes, phrases, key words, and even pictures, but the speaker must feel comfortable with the method and should use and refine—expand if necessary—the notes in rehearsal. Speech notes are not a substitute for speaker preparation.

"I . . . uh . . . forgot what I was going to say." Has that ever happened to you? If so, you'll probably agree that it's the "lonesomest feeling in the world" to have your mind go blank in the middle of your speech and realize that you don't have any idea what you were going to say. In fact, you may find yourself not even sure you could recite your name! And that, my

*From *IEEE Transactions on Professional Communication,* Vol. PC-24, No. 3, Sept. 1981, pp. 130–132. Reprinted with permission from *The Toastmaster,* Vol. 44, No. 8, pp. 11–13.

friend, is why the making and using of speaking notes is near and dear to the heart of every Toastmaster.

When I joined Toastmasters, the subject of speaking notes was a great and very perplexing mystery to me, and the few cryptic words about notes in the "Ice Breaker" section of the *Communication and Leadership* manual in use at the time didn't help too much. After wrestling with the problem on my own for a couple of speeches—without any notable success—I set about serious research on the subject, expecting to find some golden nugget of universal truth about how the "old masters" handle their notes. But guess what? After reading a pile of "public-speaking-made-easy" books and talking to every senior Toastmaster available to me, I found as many answers as I had sources.

The universal truth

Everyone's personal technique was a little different and, even worse, many of the old Toastmasters had used more than one technique from time to time. But I did find two common threads. The main thread, and the one closest to a "universal truth" I was able to find, was that the generation of speaking notes should be an integral part of the preparation of any speech. The second was that the speaking notes that are to be taken to the lectern should also be used in rehearsal.

But more useful to me was the discovery of a dandy bunch of techniques to try while I was thrashing around to find the one that would best fit my own personal needs. The most useful of these were

- The "one piece of paper" method;
- The "notes from manuscript" method;
- The "stack of cards" method;
- The "one itty-bitty card" method;
- The "barefoot in the wild woods" method;
- The "visual aids as notes" method;
- Some unique methods purportedly used by Mark Twain.

As I describe these methods, you'll see how each fits into the preparation and rehearsal of any speech.

One piece of paper

When using this method, make your first outline of the speech on a single 8.5-inch × 11-inch sheet of paper that has been divided into a convenient number of sections (perhaps four). Allow the first section for the introduction, the last section for the conclusion, and the remaining one or more sections for the body. Then stick with that same piece of paper through the full preparation and rehearsal cycles, making any other required notes on it. Finally, take that same piece of paper with you to the lectern.

The key ingredient of this technique is that you really get to know that piece of paper, and even though it may become smudged, curled, tattered—or even tear-stained—it will be an old friend there to comfort you in your hour of need. You'll know instinctively where everything is located on the page so that if you need to check your notes while speaking, your eyes will automatically glide to the right part of the page—a most comforting technique, particularly for a beginner.

Notes from manuscript

Many people who write professionally are more comfortable making their thoughts flow on paper than verbally or mentally. This technique is ideal for them and possibly for you, because it starts with the generation of a manuscript that says approximately what they or you want to say (after making an outline from which to write, of course).

Once the manuscript is finished, immediately start to rehearse the talk—without looking at the text. When you come to a point at which your memory fails, look at the manuscript to refresh your memory but mark with a red pencil the point at which you stumbled. Then, go on to the next stumbling point, and so on until you come to the end of the talk.

After the first time through, write the first word after each stumbling point on a separate piece of paper. Then try to go back through the talk using this list of stumble words as notes. If you stumble at additional places, add them to the list until you can get all the way through the talk using only this list of words as notes. Naturally, you'll not use the same words each time through because you're not memorizing the speech, only *learning* it. While you are rehearsing, you will also be developing an ideal set of speaking notes—containing only the key words you need to tweak your memory during the actual presentation of the talk.

Stack of cards

An often-mentioned technique involves the use of a few 3-inch × 5-inch cards. In this technique, outline your talk by placing each major point on a single card. Use additional cards to note specific facts or examples you intend to include.

While organizing your thoughts, you can shuffle the cards to place the points and examples in just the right relationship and can conveniently add, delete, or change points or examples at any time during the preparation or rehearsal process. Then take the final stack of cards along to the lectern as speaking notes and flip them over as you go through the talk. If you have written in large, clear letters, you can quickly grasp the points on each card and you'll know exactly where you are in the talk.

Although this is a fine technique, particularly for a beginning speaker, it has two drawbacks: First, your audience will notice that you are flipping the cards; and second, the order of those cards is rather critical. Let me tell you about the time I forgot to put a rubber band around my stack of cards and dropped them on the way to the lectern. . . .

One itty-bitty card

As your confidence increases, try reducing your piece of paper or stack of cards to a single 3 × 5 card. Once you have mastered this technique, you can slip that card into your pocket and then, unobtrusively, sneak it onto the lectern. This way, if your rehearsal is adequate to keep you from glancing down at your notes too often, you will seem to your audience to be speaking without notes. Even if someone sees you placing the card on the lectern, your audience won't be distracted by any obvious use of notes.

The key to this approach is simple, direct speech organization, because too complex an outline would require too much to be written on that tiny card (not to mention that simple, direct speech organization is better anyway!). Once your outline is ready, pick key words that express the main thought in each major section of the outline and write those words in large, dark letters on the card. Then rehearse and present your talk using that card. If you find specific trouble points in your rehearsal (for example, if you keep forgetting particularly important transition words), add a few more key words to the card. But be careful not to let the card get cluttered.

Barefoot in the wild woods

At some point in your speaking development, you'll want to leave your carefully prepared notes at home and confidently stride up to the lectern empty-handed (and with no itty-bitty card in your pocket). When this time arrives, please don't try to memorize your whole speech. I take that back; try it once like I did. You'll be cured for life.

I knew better, but the seductive idea of being able to pick each word in the talk for just the right effect was too much for me to resist—so I did it. I had worked up a beautiful talk and rehearsed it until I could get through it letter-perfect. Then, when I was introduced, I confidently stepped out in front of the lectern and started to speak. My eye contact was superb, my introduction was powerful and dramatic. But as I started into the first sentence of the body, someone shifted in his chair—catching my eye. My mind went blank. After the longest minute of my life, I finally had to walk back to my place at the table to get my manuscript. Just the comforting feel of the text in my hand was enough to get me started again, but the lesson was learned: *Never memorize a speech—memorize only a simple outline.*

If you have formed the habit of reducing your speaking notes to a few key words, you can easily memorize those, particularly if your outline is simple and direct and your organization is logical. The main difficulty, frankly, is screwing up the courage to try it the first time!

Visual aids as notes

One final technique is particularly applicable to technical and nontechnical briefings in which large amounts of information or data must be presented through visual displays. This technique uses those visuals as an outline. A properly designed set of flip charts, overhead projector transparencies, or 35-mm slides can lead you very logically through the presentation without the need for any sort of notes. However, I have found it very helpful, particularly in a long or very technical briefing, to have a list of the file numbers and titles of my slides on the lectern, so I can ask the projectionist to call up a specific slide—out of sequence, if required—to help answer a question.

Good old Mark Twain

Mark Twain, perhaps the most sought-after American public speaker in history, believed it was extremely important to speak without seeming to have notes, and is reputed to have tried several interesting techniques to that end (including writing the first letter of the first word of each outline heading on a fingernail with water-soluble ink and then licking off each letter as it was used). But his ultimate technique was to draw a crude picture illustrating the main point of each major outline topic. He found that he could easily remember a row of these pictures, and could thus carry his outline in his head.

Avoid the crutch

One "revelation" that I found over and over in my research was the admonition not to use your notes for a "crutch," which frankly caused me more consternation than comfort until I figured out what that meant. The light finally dawned when I recalled my very first public speaking experience. It was "Youth Sunday" in our little country church and I was to deliver part of the sermon. I had written out my five-minute talk word for word and didn't bother to practice because I had intended to read it. But I had been advised by my father (an early Toastmaster) that it was good form to look up at the congregation from time to time even though I was reading.

After the first line of my speech, I looked up at the congregation per instructions. The sight of 50 pairs of eyes caused a great glacier to form in my lower intestinal tract. I instantly dropped my eyes to the text and, after finally finding my place again, mumbled the rest of the talk into the page in four minutes flat. Now *that* was definitely using notes as a crutch!

My lesson, dramatically learned, was that no set of notes is ever a substitute for proper preparation and rehearsal of a speech. It's far better to avoid the crippling effects of inadequate preparation than to limp along leaning on a pile of paper.

The ultimate truth

So there you have it—the "ultimate truth" in note-making. Everyone does it a different way, and often changes methods for different types of talks as his or her experience level changes. I have tried each of these techniques and a number of combinations, but like the wise cooks who always try out a new recipe on their own families before preparing it for company, I have tried them first in the friendly atmosphere of my own Toastmasters club meetings.

By the time my sampling period was over, I had settled on the "one itty-bitty card" method as best for me, and I kept that until the time I was ready to memorize those few key words (and pictures) and go "barefoot in the wild woods." But that's just my solution for *me*; it may not be the best for you.

If you're a beginning speaker who is casting around for the ideal note-making technique—or if you're faced with a new kind of talk which doesn't seem to fit your old tried-and-true method—why not try a few of these techniques in the safety and comfort of your own club? Maybe one will be just right to help you avoid that "lonesomest feeling in the world."

C

Presentation Evaluation Forms

THIS APPENDIX CONTAINS a set of evaluation forms for the speaking exercises of Appendix A. The forms are intended for use in a technical speaking class to provide constructive feedback to the speaker. The instructor or another student in the class should complete the appropriate form during and immediately following the presentation being evaluated.

Evaluation guide for the work review

1. **Objectives**

   What action did the speaker apparently want the audience to take?

   What will the audience remember two hours after the talk?

   -
   -
   -

2. **Content and organization**

   Was the subject limited to material required to fulfill the objective?

   Were a clear introduction, body, and conclusion present?

   Was material organized to give the audience the clearest possible picture of the activities being reported?

3. **Technical level**

   Did the speaker properly define and use the vocabulary of the profession?

   Did the speaker appear to have a good grasp of the activity in his or her task or organization over the reporting period?

   Did the speaker properly explain technical concepts without "snowing" the audience?

   (No simple "yes or no" answers please—give the speaker some specific clues for improvement.)

Evaluation guide for the marketing presentation

1. Did the speaker make the sale? Why or why not?

2. Did the speaker focus on benefits rather than features?

3. Were the technical detail level and vocabulary of the talk appropriate?

4. Were the visual aids effective? Could they be easily read?

5. Were numbers properly presented?

6. Were shortcomings and competition dealt with in a positive manner?

7. Look at the notes you made during the talk. Do they support the sale?

(Please make detailed comments with illustrative examples from the talk.)

Evaluation guide for the technical conference paper

1. Was the information in the paper completely accurate?

2. Did the speaker have proper backup for controversial subject areas?

3. Were the technical detail level and vocabulary of the talk appropriate to a sophisticated audience?

4. Were the visual aids effective? Could they be easily read?

5. Was the tone of the paper appropriate to someone working at the state-of-the-art level in the field?

6. What three things do you remember from the talk? Do they support the speaker's objective?

(Please make detailed comments with illustrative examples from the talk.)

Evaluation guide for the after-dinner technical talk

1. Was the presentation both informative and entertaining?

2. Did the speaker avoid controversy?

3. Were the technical detail level and vocabulary of the talk appropriate to a technically qualified audience that has just finished dinner?

4. Were the visual aids effective? Could they be easily read in the restaurant situation?

5. Did the speaker establish himself or herself as an "insider" to the audience's professional orientation?

6. What three things do you remember from the talk? Do they support the speaker's objective?

(Please make detailed comments with illustrative examples from the talk.)

Evaluation guide for the multiple-speaker briefing

1. What was the overall objective for the set of briefings? Was the objective met?

2. Was the subject area properly divided into individual papers?

3. Did the opening remarks set the proper tone for the day and let the audience know what to expect?

4. Did the closing remarks summarize the main points made during the day and focus the audience's attention on the overall objective of the day of briefings?

5. Were the written guidelines to the speakers adequate and appropriate?

6. Were the introductions effective? Did they establish the speakers' credentials?

(Please make detailed comments with illustrative examples from the talk.)

Evaluation guide for reading a prepared paper

1. Was the paper, as read by the speaker, interesting to the audience?

2. Did the speaker achieve good eye contact while reading the paper?

3. Was the speaker able to read the paper so that it sounded spontaneous?

4. Were the visual aids effective? Could they be easily read?

5. Were the visual aids appropriately displayed during the reading of the paper?

(Please make detailed comments with illustrative examples from the talk.)

Evaluation guide for convincing the skeptical audience

1. Was the audience convinced?

2. Did the speaker have convincing, objective backup for all assertions contrary to the audience's previously held views?

3. Were the technical detail level and vocabulary of the talk appropriate to the audience?

4. Were the visual aids effective? Could they be easily read?

5. Was the author able to keep the controversy on an objective, technical level without being personally threatening to the audience?

6. What three things do you remember from the talk? Do they support the speaker's objective?

(Please make detailed comments with illustrative examples from the talk.)

Bibliography

Calero, H., and G. Nierenberg, *How to Read a Person Like a Book,* New York: Simon and Schuster, 1971.

Dlott, A. S., *Preparing and Delivering a Presentation,* La Selva Beach, CA: Communication Systems Co.

Kodak, *Planning and Producing Slide Programs,* Rochester, NY: Motion Picture and Audiovisual Markets Division.

Minnesota Mining and Manufacturing Company (3M), *How to Prepare and Present Better Meeting Graphics,* St. Paul, MN: Audio-Visual Division/3M.

Molloy, J., *Dress for Success,* New York: Warner Books, 1975.

Toastmasters International, *Humor Handbook,* Santa Ana, CA: Toastmasters International.

Toastmasters International, *Technical Presentations: A Toastmasters International Program,* Santa Ana, CA: Toastmasters International, 1984.

Vogel, D. R., G. W. Dicksen, and J. A. Legmen, *Persuasion and the Role of Visual Presentation Support: The UM/3M Study,* unpublished paper, St. Paul, MN: 3M.

About the Author

David L. Adamy is a technical professional with two electrical engineering degrees, and is also a very experienced public speaker on both technical and nontechnical subjects. He has been a military instructor, and a project manager in the military electronics industry, required to make hundreds of project reviews and other briefings (some in English and some in German, which he learned as an adult). For the past 15 years, he has made his living as a technical and management consultant, a writer, and a speaker. He has presented many professional papers to audiences of a few dozen to several thousand people. He has developed and presented dozens of short courses (one to five days) ranging from technical briefing technique to proposal preparation to hard-core technical subject matter.

Dave is also a prolific after-dinner speaker and a tough competitive speaker in Toastmasters International contests. He manages and chairs yearly technical sessions involving multiple speakers on multiple days, and manages short courses by other professionals to international audiences. He has published over 100 magazine articles, which are mostly technical, and has five books in print.

He describes himself as "an OK engineer, but a truly great fly fisherman." He has been married to the same long-suffering lady for 40 years, and has four daughters and six grandchildren. His sense of humor, although it often gets him into trouble, enlivens even his most technical writing and speaking.

Index

A

D

E

N

O

Q

R

S

T

V

W

Z

Recent Titles in the Artech House Technology Management and Professional Development Library

Bruce Elbert, Series Editor

Critical Chain Project Management, Lawrence P. Leach

Decision Making for Technology Executives: Using Multiple Perspectives to Improve Performance, Harold A. Linstone

Designing the Networked Enterprise, Igor Hawryszkiewycz

The Entrepreneurial Engineer: Starting Your Own High-Tech Company, R. Wayne Fields

Evaluation of R&D Processes: Effectiveness Through Measurements, Lynn W. Ellis

Introduction to Information-Based High-Tech Services, Eric Viardot

Introduction to Innovation and Technology Transfer, Ian Cooke and Paul Mayes

Managing Engineers and Technical Employees: How to Attract, Motivate, and Retain Excellent People, Douglas M. Soat

Managing Virtual Teams: Practical Techniques for High-Technology Project Managers, Martha Haywood

The New High-Tech Manager: Six Rules for Success in Changing Times, Kenneth Durham and Bruce Kennedy

Preparing and Delivering Effective Technical Presentations, Second Edition, David Adamy

Reengineering Yourself and Your Company: From Engineer to Manager to Leader, Howard Eisner

Successful Marketing Strategy for High-Tech Firms, Second Edition, Eric Viardot

Successful Proposal Strategies for Small Businesses: Winning Government, Private Sector, and International Contracts, Second Edition, Robert S. Frey